LIFTING THE FLOOR

Revealed: the true stories hiding beneath the tiles of the data centre industry

MICHAEL TOBIN OBE

TOBIN VENTURES

TESTIMONIALS

'This is a "top-shelf" canter along the information super-highway to the secretive hubs of data-storage. Those that grasp all the lessons will be the winners.'

ALASTAIR STEWART OBE

'Michael Tobin is one of Britain's most successful and inspirational business leaders.'

GAVIN ESLER

'In true Mike Tobin style, an honest, riveting, witty, and enthralling "behind-the-scenes" extraordinary journey into the world of data centres, boardroom manoeuvres and corporate deal-making.'

VINAY NAGPAL, PRESIDENT, INTERGLOBIX

'An inspirational read. Michael proves no challenge is ever too big. If you want something, you can achieve it – if you're prepared to listen, learn and work hard.'

ALEC STEWART OBE, DIRECTOR OF CRICKET, SURREY CCC

'Who else would be better to tell the insider's story than the father of the UK Data Centre Industry, Mike Tobin? Who took his company from near bankruptcy to a $2 billion+ public business!'

JOHN O'CONNELL, CHAIRMAN, SCALEUP GROUP

'Over the last couple of decades, I've come to respect Mike as a wily competitor, combining a street-fighter like tenacity with a passion to win. "Lifting the floor" is a fascinating "ring-side" account of his role as one of a handful of leaders that dragged a failing, fledgling industry into a multi-billion-dollar global phenomenon.'

JOSH JOSHI, CHIEF FINANCIAL OFFICER, INTERXION, 2007-18, AND TELECITY, 2003-6

PERSONAL DEDICATION

Success is never permanent, just as failure is rarely fatal.
Success is more like a journey than a destination;
you need support along that journey.

My journey has only been possible because of my wife Shalina;
my strength, my shelter, my life, my love.

CONTENTS

ABOUT THE AUTHOR

Born in the backstreets of Bermondsey, serial technology entrepreneur Michael Tobin OBE is known as the 'Maverick'[1] former CEO of the Telecity Group. He is also known for his unconventional management style, like the time he took his team swimming with sharks to teach them how to manage their fear. Today, he is widely credited with having been *'instrumental in creating the digital infrastructure of the internet of Europe'* (Ed Vaizey MP).

In 2014 he was recognised by Her Majesty the Queen with an OBE for Services to the Digital Economy, following his unwavering dedication to the data centre industry. His day at the Palace was a long way from his humble childhood beginnings that included suffering periods of homelessness, violence, and dodging petrol bombs in Rhodesia.

Michael's outstanding achievements earned him many awards during his career, including 'Datacentre and Cloud Influencer of the Decade' (Broadgroup Industry Awards), 'Top 25 Power Individuals of Industry' (Smith and Williamson), 'UK IT Services

Entrepreneur of the Year' (Ernst & Young) three years running and 'Lifetime Achievement for Services to the Data Centre Industry' (Data Centre Europe Awards).

Today, Michael works around the clock in his Non-Exec Director and Chairman roles and undertakes charity missions that continue to test his limits. His latest missions, which support and empower young and vulnerable people, have included running 40 marathons in 40 days for the Prince's Trust and undertaking a challenging and dangerous trek to the South Pole in January 2020, supporting The Brain Tumour Charity, to beat brain tumours in children.

EVERY SAINT HAS A PAST AND EVERY SINNER HAS A FUTURE.

THE CONNOR BROTHERS

WHO IS MIKE TOBIN?

Not a man you forget

I was at a party in Vegas…

As luck would have it, I entered the party and probably within the next 60 seconds, this very special guest was entering the party too. I was told to introduce myself to him and welcome him. And so I had the honour of doing that…

But as I handed over my simple business card to the special guest, what I got in return was a stack of business cards in my hand. I looked at it and wondered, what is this thing? It's literally five to six business cards that are stacked together. So I started looking at them. One says Mike, the next one says, Tobin, the third one has half of Mike's face on it, the fourth one has his number, the fifth one has his email… and so on. Talk about being innovative and leaving an impression. Obviously, I'd never seen a business card like that. Sure enough, that special guest was Mike Tobin.

Here he was in in his cowboy boots, ripped jeans, shirt, a couple of buttons open. And I just started wondering, is he a customer? Who is this guy? One thing was for sure, I wouldn't forget him!'

VINAY NAGPAL

FORMER VICE PRESIDENT, PRODUCT MANAGEMENT AT DUPONT FABROS TECHNOLOGY

The guy in the ripped jeans

Vinay is right, I made sure people didn't forget me. The fact that he still remembers a business card keyring I'd had made over a decade ago says a lot. I had all sorts of other versions of business cards too, which became important talking points and helped me start new business relationships.

The funny thing is that people always mention my unbuttoned shirts now too – but little did they know there was purpose in my personal presentation. When we were doing our public company results, all the analysts, investors and institutions would come in, and the first thing they'd look for, before anyone even started talking, was how many buttons I'd undone. If I had three or four buttons undone, it was going to be brilliant results. If I had two buttons undone, they were just going to be good results. So that was the first indicator whether the announcement was going to be good or great.

Many years ago, I was trying to get back to London for a data centre industry awards ceremony hosted by Philip Low of Broad Group. I had been nominated as Personality of the Year, and the organisers were very keen for me to be there. But I had taken a few days off to go skiing in Switzerland with my kids and was planning to fly back for the ceremony. On the day of the event, the weather was against me and all the flights were delayed or cancelled. I phoned in to say it didn't look like I was going to be able to make it after all. 'Please make it', was the message back. 'We really need you to be here.'

There was a break in the weather, eventually, and – well behind schedule – a flight managed to get off the ground. There was no

I made sure people didn't forget me

time to go home and get changed into black tie, but I thought that if I went straight to the function, I might just get there in time. I landed in London, and with minutes to spare a cab whisked me over to the Grosvenor House Hotel, just in the nick of time. As I walked into the ballroom, which was full to bursting with industry colleagues in immaculate tuxedos and evening wear, heads turned as they clocked my T-shirt and ripped jeans. 'It's black tie,' hissed someone helpfully. I said, 'I know, I know, I couldn't help it,' and just as I replied, my name was called out to go and collect my award.

It was the main award of the night, which obviously explained why the organisers had been so insistent on my attendance. I made my way to the podium, where the comedian Ed Byrne, who was compering the event, made a suitably sarcastic comment about my choice of clothing. 'You could at least have made a fucking effort,' he said, 'Hark who's talking!' I replied, since he was looking pretty scruffy too. We all laughed, and the evening moved on. But the impact of that night lasted a long time. I was from then on remembered as the guy in the ripped jeans who had picked up the main gong.

I went on to win that same award for the next two years running. On the last occasion, I actually said, 'Right come on then, someone else needs to win this next year!' But as I looked at the marble base of the award, with my name engraved on the silver plinth, I realised my industry reputation was now truly set in stone.

FOREWORD

I first encountered Mike across a very large conference table at a meeting hosted by a major investment firm where my colleague Steve Wallage had been invited to comment on the data centre market. As those who might have experienced these types of meeting are aware, the atmospherics are not exactly relaxed. But Mike's demeanour despatched any evidence of tension as he sat leaning back, smiling frequently throughout the questions. His confidence and demeanour have become his own trademarks, or personal brand if you like, and one that has carried him to not one but many successes and achievements in his career.

He first appeared at an awards ceremony we had organised in 2007, just returning as he had from a ski trip. As most of us who know him are aware, his style of dress has sometimes been what we can describe as adventurous. Yet this also betrays his attitude towards conformity, and how ultimately, it is the trust you have in yourself and exude to others that really matters.

But from my own perspective, as an event producer, he was and is an industry star turn. He always speaks very directly and frankly and is completely unafraid of challenging questions – and asking them of others.

He remarks in this book on how he learned to detach himself from fear and worry. But seeing him in action, there is also more than a hint of relishing confrontation. We have engaged him many times as an expert speaker and enjoyed over the years his feisty exchanges, with journalists and investors often in the firing line.

Above all, we came to know him as the friend he became to us. And one with a tremendous grasp of businesses in the sector, and in recent years, a more formidable reputation as a deal maker and astute investor.

This new and truly fascinating insight is about himself – perhaps few know of his years of experience living and working across several cities in Europe – as much as about the business of data centres, and riveting accounts of the sometimes extraordinary and bizarre if not scurrilous people he has had to contend with on his journey. This compelling story evidences his sheer resilience, focus and risk-taking in driving a company through some fairly hairy times.

As with all things in life, luck sometimes deals a hand, and insurance money for the unfortunate flooding of a newly built data centre in Prague saved the nascent Redbus business from sinking and allowed him the funds and opportunity to survive many more twists and turns into what eventually emerged as Telecity. He cites it as a key turning point.

Many reading this book will find names very familiar to us in the data centre industry. But it is as enjoyable as it is gripping in its account, and lifts the lid as much as the floor, with a gossipy rendition of the jets, hotels and boardroom battles involved in building what became a multi-billion-dollar public company, and is pleasingly littered with frank opinions of the many people (some may not be quite as pleased!) he encountered on the way.

Nevertheless, as with all entrepreneurs, the story told here tells of a singular bottom line ability common to them all: guts.

PHILIP LOW
CHAIRMAN, BROADGROUP

INTRODUCTION

Can you imagine your life today without the internet?

Getting online is as important to us now as getting around was when the wheel was invented over 5,000 years ago. Luckily, in this day and age, if my wheels stop working I can usually find an alternative way to get around. I can think of at least one occasion in the last few years when I haven't been able to get online at a crucial moment, and I've gone to all sorts of extreme lengths to get connected. A few years ago, I was travelling between New York and Amsterdam via London, and when transferring at Heathrow I got off the plane and realised I had left my wallet and my phone on the plane in the little drawer next to the seat.

I went to the ticket desk and implored them to please 'Find my phone!' I was so desperate, I even said the cleaners could keep the wallet. I missed my connecting flight to Amsterdam while I waited for them to find my phone. It wasn't the phone calls I was worried about, it was needing to get online and get my emails sent!

Years ago, I used to visit a tiny village in the north of Portugal called Prova with the family, and the only place you could get a network signal was on top of a massive rock just outside the village, called Fragao de Pendaun. I climbed that rock every single day just to send and receive emails while I was on holiday.

Yes, it's incredible that taking away internet access can, at times, quite literally bring a grown man to his knees. But thanks to our ever-growing reliance on the net to work, travel, shop and communicate, data centres, (the physical home of the internet) now hold the keys to our lives.

Today, thanks to the catastrophic impact of the COVID-19 virus, the whole world has experienced a terrifying reminder of how vital the internet is to all of our lives. Since the end of 2019, when the virus first struck in Wuhan, China, most of the world has been in lockdown. In the race to save lives, every single office, shop, restaurant, bar, hotel, leisure centre, play area and non-essential public place was closed. Police patrolled the streets across the world, ensuring that people stayed in their homes to slow the spread of the potentially fatal disease and put an end to what felt like a biological apocalypse.

Everyone who could work from home was told they must. Every child who could stay away from school was educated at home. Only our most essential key workers and frontline medical staff could leave their homes to help save lives as the death toll rose daily. The streets were empty and the world economy was in crisis.

This was Armageddon for the internet. Literally every professional or social activity that could go online did, and the global infrastructure of the worldwide web felt the strain. Whilst the world ramped up online streaming, and virtual socialising, working, and communicating digitally became the new normal, a series of continuity measures to 'mitigate the higher demand for bandwidth during the quarantine period[2]' were put in place. For the first time ever, the big guns of the internet like Netflix,

YouTube and Amazon had to pull together to help maintain efficiency, supporting the net to keep running smoothly, and ultimately keep the world turning.

The internet is now as essential to our existence as electricity and gas. What this means for the future as to how we use the internet remains to be seen. But one thing's for sure, right now, to maintain any degree of normality, it is our only hope.

What is a data centre?

If you've never seen a data centre, imagine a large room with racks and racks of computer servers, all stacked up and whirring away whilst they work hard to process, save, exchange, back-up and recover data. Together, they generate a huge amount of heat whilst they are busy storing websites, running email, social media and instant messaging services, enabling e-commerce transactions, powering-up online gaming communities, supporting communications platforms and so much more.

So how did the data centre industry begin?

Why are data centres so important to our lives? And what would happen if they stopped giving us access to the internet?

1990 was the start of a new era. Engineer and computer scientist, Tim Berners-Lee, had created the World Wide Web in 1989, and less than a year later, quietly released it into all of our hands. The internet soon became one of the biggest game-changing inventions the world has ever known, but at the time, few people

"

There are now
over 8 million
data centres
in operation
across
the world

"

knew what use it would be to them, and even fewer had the tools or ability to use it.

Meanwhile, more and more of us were using personal computers, and large businesses were using more computers in the workplace. Many had even created their own unique networked 'data rooms' to increase their speed and processing capabilities. This was essentially their own small version of a data centre, which allowed them to store and access large quantities of their data themselves, but it had its limitations.

The in-house data room was not a practical or realistic solution for many businesses, and larger, external data centres were quickly needed to be able to handle the sheer volume and scale of traffic, data and communications now travelling via the internet.

The first true data centres were born out of the dot-com bubble between 1997-2000, when companies demanded faster internet connectivity and continuous operational support to start, grow and maintain their presence online.

Since their inception, data centres have not only become essential for all of us to fulfil our daily needs, they have created their own economy, industry and community. There are now over 8 million data centres in operation across the world. The largest individual data centre is owned by Microsoft, with well-known players like Amazon, Google and Facebook following suit. The Data Economy is valued at a staggering $3 trillion globally and growing exponentially.[3] It all needs to live in data centres…

How important are data centres?

Almost every business in the world, right down to your local convenience store, will use data centres in some form or another, whether that be for housing its website, or enabling online purchasing from suppliers. Even when you pay for bread in a supermarket, your credit card details are taken and whizz through the air to a data centre in seconds. In many homes our communication (telephones), entertainment (TVs, game systems, speaker systems), electrical appliances and even security systems are connected to the internet and rely on multiple relationships happening seamlessly inside a data centre somewhere.

Although they have only relatively recently come into existence, these centralised hubs of data processing are now essential to pretty much every business and individual in the world, and without them, business stops, and chaos ensues!

It's hard to believe the volume of business a data centre supports or, conversely, if it fails, the volume of business a single data centre can bring down. So, let me give you an example. Do you remember the British Airways crisis in 2017 that grounded 459 flights on a busy holiday weekend, resulting in chaos, anger and distress for 75,000 stranded passengers?

When the story broke, I was interviewed by Sky News, who wanted to know how or why BA's Heathrow-based data centre could have possibly 'failed' – leaving such a trail of disaster in its wake. I explained that outages happen, but are extremely rare in the data centre world, and when they do, it's usually down to human error – one way or another. And I was right.

BA had tried to claim that a 'Power surge on the National Power Grid' was to blame[4] in an attempt to divert responsibility. It turned out though that, in simple terms, when a problem had occurred within BA's computing infrastructure, someone had decided to do the old 'turn it off, turn it on again' trick. Except what they clearly didn't realise was that when they flicked it back on and powered up the entire bank of servers at BA's data centre simultaneously, this would create a power demand so huge that it would frazzle the system by pulling too much current at start-up! It wasn't the Grid, but human error that was to blame.

Let's just say that alongside the very unhappy customers who missed their flights, the well-intentioned and wonderful BA staff who had to sort it all out, the reputational damage, and the hundreds of millions of pounds it cost BA shareholders in lost profits, it wasn't the best day for Willie Walsh, BA's CEO, or for the engineer who flicked the switch!

Lifting the floor

During my time leading such a fast-growing, but turbulent industry, I was involved in some of the most intense, unthinkable, and often life-changing situations that took us from the creation of the UK's leading data centre through to the expansion of the entire industry across Europe, the US, Africa and beyond.

This is the never-before published, true story, behind the birth of the data centre industry in Europe, which took me from being a Sales Director to receiving an OBE for my Services to the Digital Economy. In putting this book together, I have talked to former rivals and industry comrades who have either had a fundamental impact on the industry themselves, or experienced the highs and lows of being part of an emerging industry over the last 25 years. Together we reveal the hidden stories of the events and the people behind the data centres who found themselves caught in a web of a very different kind.

Read on to discover the scandalous stories and real boardroom battles that have been lying dormant until now in this explosive book, which 'lifts the floor tiles' of the industry to uncover the chaos lying below.

PART 1

THE BIRTH OF THE DATA CENTRE INDUSTRY:

What was I getting myself into?

GETTING ON BOARD THE BUS

The dot-com bubble

On the 10th March 2000, just a few months into the new millennium, the dot-com bubble finally burst. The historic tech boom that had started in 1995 ended as quickly as it began.

For five incredible years, between 1995 – 2000, the world had been swept up in a dot-com frenzy. Computers had become a necessity instead of a luxury, connectivity had improved dramatically, and people were spending more and more time at their personal computers; working, shopping, trading, playing and communicating online.

At the same time, spending was up – interest rates had dropped lower than they'd been in the past couple of decades, which meant that capital was readily available, and borrowing money was easier than ever before.

Over this period, investment in technology was being heavily encouraged by the banks, who naively assumed that any new internet company wielding the .com suffix was bound for a bright future. In all this excitement, the world had become blind to the traditional formulas of establishing a company's value, and confidence in dot-com companies was so extreme that it was possible for an up-and-coming new business to raise a significant sum before it had even generated any revenue, let alone turn a profit – all based on a shaky promise of online success. In reality, these companies were, of course, often massively overvalued by hundreds of multiples and would never ever become profitable. The mountain was just far too steep to climb for most, who were simply acquiring more and more debt. Indeed, we still see this to a lesser extent with businesses today, such as Uber, who – despite being incredibly popular – have themselves admitted that they may NEVER turn a profit[5]!

There had also been a huge amount of personal investment happening during the 90s, with 'ordinary' people jumping on the day trading bandwagon, leaving their stable jobs to make a fast buck by buying and selling stocks and shares like they were the next Wolf of Wall Street. Of course, the media fed the public's hunger and desire for this exciting new economy; reporting on its movements with all the glamour and pizazz of a Super Bowl game. But, as the saying goes, 'what goes up must come down' and when the dot-com bubble finally burst, it dropped everyone in it back down to earth with an almighty thud.

During the boom, I was working and living in Europe. In 1987, I started an 11-year stint in France, developing international technology businesses across Europe, including amongst others, Computacenter's international operation, ICG, and a US super-server manufacturer called Tricord Systems. I then moved to Copenhagen in 1998 to work for ICL, a large British computer mainframe and services business. ICL was the UK's answer to IBM. I was there to turn their mainframe maintenance service and repair company into a managed outsourcing provider (providing a more comprehensive suite of technology services to customers). Mainframes were on the way out in favour of Personal Computers. The company had already stopped selling new mainframes and had simply become a repair centre for the existing installed base, so it was a declining business and had to evolve its offering or it would eventually disappear. Copenhagen was a great city and my daughter was born there in November 1999, giving me fond memories of that time. The IT industry was going from strength to strength, and so was I.

The turnaround of ICL Denmark was as stark as it was rapid and, in 1999, in recognition of my contribution, I was selected from around 20,000 ICL employees around the world to take part in their Millennium Programme, joining the 'Top 25 stars of the future from around the world' in a project that would prepare us for potential leadership and corporate board roles in the company's future.

I'll never forget the 18 months we spent together on the programme, being challenged in every way possible and learning significant life lessons from all kinds of teachers, including a psychologist and a yogi. The unusual management practices we learned have stayed with me to this day and proved both extremely valuable and entertaining for my employees and colleagues ever since. For example, over 15 years later, when I was at Telecity, I brought Jagdish Parikh, a Harvard MBA who is also a business yogi, into the office every year to sit with my management team for a day and teach them how to hypnotise themselves. He showed us how to become detached from our physical self, to de-stress, to sleep well, and not to give in to fear and worry but how to live in the present instead and stay focused. All lessons I still live by today and which have led to my reputation of doing things very differently.

Life wasn't all a fun adventure at ICL. There was the takeover bid from Fujitsu, the world's second-oldest IT company after IBM. ICL had been changing significantly during the 1990s, and the Japanese had set their sights on acquiring us – which they soon did. Eventually, in 1999, Fujitsu (now including ICL) merged with Siemens to become *Fujitsu Siemens Computers*. The evolution of ICL from mainframe manufacturer to PC manufacturer was complete, and by 2001, I was working for what was now the fifth-largest computer manufacturing company in the world, and it was a great opportunity to make my mark on their future.

At first, my mission as managing director at ICL Denmark had been to help turn the loss-making company around, which we did – taking it from losing millions each year to profitability in just two years. It had been challenging, but it wasn't my first MD role. I had run businesses in multiple countries around the world and, ultimately, my previous successes, along with my international experience, led me to being offered this new role. I was then asked to move to Frankfurt and set up Fujitsu's new e-business division's German subsidiary. I completed this task with great success, taking it from the start-up phase to a company employing 80 people within a year, and achieving revenues of $10m by its second year.

Turning tides

We had been riding high on the tide of a new digital age, and I had made a good life in Europe over the years, but by 2001 my world had changed dramatically. I had become a father. My daughter Eloise was born in Denmark two years earlier, and we had another baby on the way. Meanwhile, businesses were facing an uphill battle as the negative impact of the dot-com crash spread across the globe in heavy, turbulent waves. Then, to everyone's horror, the whole world was sent reeling by the devastating local, national and international impact of 9/11.

On the 11th September 2001, I was based in Frankfurt, working for Fujitsu when the planes hit the Twin Towers. A moment I will never forget. I remember trying to watch the news footage live on the CNN website, which had crashed through the sheer volume of people trying to access it. On top of the shock and obvious concern for people's lives, I remember thinking that the infrastructure of the internet clearly couldn't support mass access!

Exactly one month after the horrific attack brought New York to its knees, my son Nelson was born in Germany.

I had always been inspired by the images of the East End of London in ruins during the Blitz of the Second World War, all except for St Paul's that is, which stood tall amongst the flames and debris. I then learnt that right at Ground Zero in New York, there is a tiny church, also called St Paul's, that by all rights should have been destroyed by the falling Twin Towers.

Miraculously, on that fateful day, a sycamore tree fell against the church and somehow protected it from all the debris. The church became a safe haven for the families and friends of the victims; somewhere they could go to pray and find some degree of peace and comfort. And it was not the first time that the church had survived disaster. In 1776, the church had withstood The Great Fire of New York City. As a result, it became known as The Little Chapel That Stood. It captured my heart so much that we gave Nelson the middle name Paul to represent this resilience and strength.

As they grew up, I told my children the story of the little church, and how it had been so strong. Years later, I took them to New York to let them see it for themselves. There, in front of the church, was a bronze cast of the tree that had protected it. In a strange twist of fate, and to my great surprise, when we read the plaque, we found that the artist's name is Steve Tobin. It was a profound moment seeing that we had the same surname!

Meanwhile, across the US, the data centre (or as they say, 'data center') industry had started to thrive in the new digital age. The most renowned company being Equinix, which was founded in California in 1998 by the late Al Avery and Jay Adelson, who had originally worked together as facilities managers at Digital Equipment Corporation. Jay, in particular, was considered a young internet entrepreneur, and just 10 years later was named as one of *Time* magazine's Top 100 Most Influential People in the World.

Al and Jay's vision was to create a Neutral Internet Exchange (NIX), which is the physical place in which critical information could be exchanged. Think of a major international airport, such as JFK, Heathrow, Schiphol, Dubai or Singapore. These airports are some of the largest in the world, not just because of the local demand, but because they have become transit hubs for travellers from all over the world. Just as these airports serve as airline traffic hubs, Internet Exchanges serve as data traffic hubs.

A tremendous amount of data moves around whenever we do anything on the internet. Even a simple Google search generates a ton of movement. Imagine you search for a hotel on Google. The information you type into your computer has to fly down the lines and find the Google page, which may be on the other side of the globe. Google then processes that request and sends out thousands of requests to millions of different websites and the information is returned to Google, and those results are then sent to your computer. A simple search for "Hotel" will return more than 8.3 BILLION results! All of those results will present themselves to you as links on your screen, and then you continue your digital journey. Imagine if this had to happen with a direct link between every single computer and every single website on the planet! That's where the Internet Exchange comes in. Everyone is connected to the exchange, and therefore to everyone else. That search today, with over 8 billion results, will take less than one second.

This move by Al and Jay, as it transpired, was instrumental in helping the entire internet to grow across the world. Possibly one of the MOST influential actions in the entire growth of the internet, and certainly one of the least understood. By 2002, Equinix had expanded to Asia-Pacific, and in 2007, it expanded into Europe and has continued its global acquisitions and expansion ever since. But more on this later...

A change is as good as a rest. Or is it?

I don't know if it was a result of the seismic shifts going on in the world around me, or some other driving force, but in February 2002 I decided it was time for a major change in my life. I had spent the last two years trying to limit the damage of the dot-com crash on Fujitsu, but enough was enough, the world had changed and so had I. It was time to come home.

So back to the UK we came, with all the hopes and dreams that come with a new start. I was transferred by ICL Fujitsu to their Windsor operations because they wanted me to carry on working with them in some undefined capacity in the UK. But now, back on home turf, 15 years after I had originally left for France, the landscape had changed, and I felt I needed a new plan.

I had worked hard over the years to grow businesses during a new and unpredictable technological era, getting them over the many hurdles they faced. And I was proud of the reputation I had built as a transformational business leader, so I felt confident and ready for a new challenge. Little did I know that the next decision I made would not only shape my own future but also change the face of global connectivity for evermore. And if you'd told me that my next career choice would see me engage in one of the hottest boardroom battles that the technology industry had ever known, and I'd see my face splashed across the press for all the wrong reasons, I would never have believed you. But, if I'm honest, I probably wouldn't have side-stepped the challenge either. I've never really been an 'easy-life' kind of guy!

Once I was back on British soil, I started going through the recruitment process of looking for a suitable job – speaking to head-hunters and executive search agents. I had many contacts in the technology and investment industries by now too, so very quickly, I had recommendations both from private and public investors to look at the data centre industry.

I had an interview with a company called Exodus Communications based somewhere around Carnaby Street. I remember I was kept waiting by a larger than life Scotsman called Ray Sangster, who was the general manager for EMEA, and an HR guy called Derek Wetter.

I remember thinking, 'Blimey, the last time I was in Carnaby Street, it was as a kid with my mum.' That was in the swinging sixties when it was the centre of the fashion universe. Now it was full of shops selling touristy tat. Was this really the headquarters of a tech company of the future…?

When he finally came into the room to meet me, Sangster basically did all the talking. In fact, I don't even think he asked me a single question about myself for a good 30 minutes, and instead extolled the virtues of data centres, and how these things would soon be the focus of our world. He told me how his company had spent millions on building some of the world's finest facilities. I thought, 'What a prick.' He was one of the most obnoxious arseholes I had ever met, but in one way he was correct; data centres were going to be massive, just not his!

I am not sure why I didn't get the job. I don't think I could have said anything wrong at the time as I was barely given a chance to speak! Perhaps I hadn't been sufficiently impressed by his 'presence of greatness'. But it was a blessing in disguise. Exodus was to declare Chapter 11 bankruptcy within seven years of opening, putting hundreds out of work. It had spent hundreds of millions of dollars to build data centres just when the dot-com bubble had burst. It had to admit defeat. Exodus was later purchased by Cable and Wireless who acquired the best 30 of their 56 data centres for about $575 million in cash.[6]

Nonetheless, something about the role these buildings had to play in the future had resonated with me. Coincidentally, very soon after this disastrous interview, I was asked to consider a role at Redbus Interhouse PLC. It was a new company listed on the London Stock Exchange, operating data centres around Europe, and headquartered in London's Docklands. I thought, 'Why not?!' It seemed to be doing well, having taken advantage of the world's growing need to be connected to the internet around-the-clock. I was told that the partners leading the business needed a sales and marketing director... and they wanted me. It didn't sound too difficult, and I'd always liked the challenge of boosting sales, so I thought I'd give it a go.

What I didn't let on though, was that I didn't actually know much about what a data centre was or how it operated. But it seemed pretty straightforward, and I'm nothing if not a fast learner. At the end of the day, with a young family to support and the potential prospect of being made redundant from ICL Fujitsu, it was important for me to lock down a new role as soon as I could. The Redbus job seemed to provide everything I was looking for, or so I thought.

Hindsight is a wonderful thing, and I now know that if I had taken some time to do my due diligence on the company, I would have discovered the 'colourful' history of the Redbus founders, and more importantly, the dire financial situation they had got themselves into, and I would never have taken the job. So this is your opportunity to find out the secrets that changed the course of history forever, and if you only take one lesson from this book, it's to *look before you leap*.

Angels and Demons

Redbus Interhouse was the joint project of two wealthy business-men, Cliff Stanford and John Porter – both of whom, it's fair to say, had already (or would soon) become infamous in their own right for their astonishing individual stories. My first meeting with the unlikely business partners probably should have raised a red flag in my head, but I was distracted by their very distinct characters, which were in equal parts intriguing and bemusing.

My first interview with Cliff was held in Canary Wharf in January 2002, in an office that unbeknown to any of us at the time, would later become mine. Cliff was a chubby, bearded, larger than life character who spoke eloquently, smoked profusely, and was fluent in Spanish. He had big ambitions for Redbus and an ego to match – but just enough charm to get away with it.

I found out that Cliff had been very lucky over the past decade; whilst many had suffered from the rise and fall of the digital economy, he had become one of the lucky ones – a 'Dot-com Millionaire'. His success had given him enough money to live a luxurious lifestyle, and enough financial security to start the Redbus Group.

Cliff was a complex person with an interesting story. As a young man, he had been fascinated by business. People said he had endless drive, and a natural eye for money-making opportunities, which is no surprise when you know that his mother, a qualified accountant, was the first woman in England to run a bookies. He was also a huge chess enthusiast, which I can only imagine helped his mental ability to make so many strategic moves later in his life.

Cliff's father left home when he was 11 years old, and he had naturally followed his mother's career path, leaving school at 16 to train as an accountant. In the early 90s, Cliff ran a small software business called *ImPETus*, which developed software for the Commodore PET (one of the first mass-market home computers of the 1970s, which looked like a small TV on top of a boxy keyboard with a cassette player built in)[7]. This was around 1992, when only large corporations and universities had access to the net, but smaller businesses like Cliff's were realising that if they wanted to survive, they needed to get online too.

Cliff became acutely aware of the growth of the internet, and it sparked an idea. He was always one to chase the dollar, so with a renewed focus on making more money, he quickly worked out how he could jump on board the internet gravy train and get rich quick. All he needed was to buy a piece of the 'internet pie', and he'd be able to sell it again for more money.

At the same time, Kent University had decided to make some money from the network boom, putting their internet access up for sale for a tidy sum of £20,000. Cliff snapped up the opportunity and turned it into the biggest money-spinner of his life. Using a *pile it high, sell it cheap* approach, he sold 'all-you-can-eat' dial-up internet access packages to 'ordinary people' for an unbeatable £9.99 per month. It was a genius move.

Remember, this was a time when we had to 'dial-up' the internet to make a connection, and it was tedious and time-consuming by today's standards. The internet didn't just appear on your computer ready for you to access anytime, anyplace, anywhere, like it does today. At the time, Google wasn't even a glint in Larry Page and Sergey Brin's eye.

To get online in the 90s, we basically had to 'call-up' information by sending a message from one modem to another, across a repurposed telephone infrastructure. This is a sight and sound that my children will never know, but it's one that I will never forget. It was the soundtrack of a whole new era. It started with a dialling tone, followed by a high-pitch whistle, which hissed and fizzled and sizzled until it merged into a kind of digital babbling brook and resulted in that eureka moment when you made a connection! The only way you'll hear it now is to visit the online Museum of Endangered Sounds[8].

Cliff's cost-saving internet dial-up service for the home-user quickly became popular, and in 1992, with 200 subscribers, he secured his place in history by creating Demon Internet – one of the first Internet Service Providers in the UK to enable people like you and me to access the net from the comfort of our own homes. Within just four years, he grew Demon from a small internet hosting business, owning just eight modems and one leased line, to a trendsetting brand with a phenomenal 50,000 subscribers connecting worldwide via more than 4000 modems.

The mark of the devil

Demon by name, and as I later found out, also 'Demon' by nature, Cliff's rise to fame was soon creating headlines that matched his behaviour to his brand name. Never one to do anything quietly, he loved the attention that his new thriving business brought, and he did whatever he could to court even more. He even changed all his phone numbers to end in 666, the mark of the devil, which was, of course, a great marketing move! You've got to give it to him, the man had vision, and the balls to make things happen, but he was a strong character and dealing with him was destined to be a rocky road.

By 1997, Demon had grown fast and Cliff thought he had made it. But in business, fast growth breeds high expectations from shareholders, and as quickly as it had become a success, Cliff's plan started to unravel. He found himself at the heart of a financial feud with Demon's investors, who wanted to oust him and take control of the business. According to *Nothing Like a Dame: The Scandals of Shirley Porter* (Andrew Hosken, 2006), in order to retain control of the business, he needed to lay his hands on £500,000 – and fast! Unable to raise the capital himself, he needed an Angel investor, who appeared before him in the shape of the renowned John Porter. This all happened well before I met Cliff or John, and it set them up as a formidable partnership that just a couple of years later would make a big impact on my life.

An unlikely partnership

A few years earlier, Cliff had been introduced to John Porter by Anthony Rothschild, an associate they had in common. John was the ideal investor for Cliff, not only was he the offspring of very wealthy stock, he was a shrewd businessman. Back in the 1980s, he'd made a $50 million profit on the sale of Verifone, the credit card payment facility company he sold to Hewlett-Packard for $1.4 billion.

John, however, came with his own interesting legacy. He was the grandson of Sir Jack Cohen (founder of the Tesco supermarkets empire) and son of Dame Shirley Porter (the former leader of Westminster City Council, and the only person since World War II to have been convicted of gerrymandering).

John was financially astute, secure and focused on investing in fast-growth businesses. Demon ticked all the boxes for him – the business had great potential for growth, and Cliff was a charming chap when he needed to be. What more could an investor want? So, in 1998, when Cliff needed an Angel to help him win the battle for control with his investors, John provided the solution to all his problems.

With John's eye for a good investment and Cliff's lack of funds and immediate need for cash, they were the perfect pair. John stepped in just in time, lending Cliff the half a million pounds he needed to keep himself at the forefront of Demon, under an agreement that gave them both exactly what they wanted out of the deal. It was a match made in heaven – or as we all later experienced – hell.

In 1998, with the security of being firmly back in the driving seat, Cliff took the bull by the horns and sold Demon to Thus, a Scottish telecommunications company, thereby pocketing a life-changing £33 million for his work, and giving John a very rapid £2 million profit on his investment.

So delighted were the pair, that after Cliff's impressive four-day celebration at his local Finchley pub, they decided to go into business together – setting up Redbus Interhouse with an investment of £2 million. The newly formed company's first colocation data centre was in London Docklands, offering web hosting and internet services to businesses. The men were now equal partners, and according to Hosken[9], John had become the UK's very own 'Dot-com Emperor' who spoke in his very own 'techy' tongue. Now he had Cliff at his side they would rule the Empire.

But their relationship was not built to last, and the boardroom battle that later ensued was to become as dramatic and bloodthirsty as any Roman conflict – and I would find myself firmly in the middle of it.

A true 'sleeping' partner

My first encounter with John Porter was in 2002, just after Cliff had given me the job of Sales Director. John had asked to meet me for an informal lunch to get to know each other and his PA,

Billie MacCarthy, had booked us a table at Smiths in Smithfields, London. Without having much time to do my due diligence on the man, I was prepared to make my judgement on face value.

He was 41 at the time. A dark-haired, good-looking chap, dressed in a smart casual way, but of obvious quality. I was told that John had a 'strong sense of urgency' about him, which I quickly discovered for myself when we met. He was a purposeful man, with the ability to skip very quickly between different topics and still come back to the first thing he was talking about. However, to my surprise, I realised that his need to deal with things quickly was because he was also a narcoleptic. Whatever he needed to do or say, he had to get out quickly before he randomly fell asleep!

I sat down and introduced myself, then John said, 'Tell me about yourself...' and promptly fell asleep in front of me! I didn't know what to do. Should I carry on like nothing had happened or try to wake him up? It wasn't your average interview, but it was something I later became accustomed to and learned to use to my advantage. I would go into his office and start telling him about something knowing he would fall asleep, then I would stop talking and wait for him to wake up a few minutes later and I'd say, 'So we are all sorted then?' and he would agree. Perfect!

Another occasion John's sleep issues came into play was when Cliff, John and Redbus CEO Kevin Neal went on an IPO road-show tour in America, despite the fact that no investors in the USA had shown any interest in them. Kevin, historically, had been Cliff's colleague at Demon Internet and was a real Essex lad. He had employed three sons in the business and smoked like a chimney. Kevin has sadly departed this world now but was a real rough diamond! They all decided they wanted to learn how to fly and all get their pilot's licence whilst they were away. They all took their partners and spent two weeks in Florida learning to fly, but for John, it was a waste of time and money.

In the end, Cliff and Kevin got their licences, but John fell asleep at the controls, so the instructors quite rightly refused to train him. So he sent for a drunk Irish pilot to come out to Florida to teach him to fly. I was told that this did the trick and John got his pilot's licence. I for one would never fly with him at the controls!!

Despite finding them both a little 'unusual', after being interviewed by both Cliff and John, I decided to take on the role of Sales and Marketing Director at Redbus Interhouse. I was on a new mission and I was looking forward to getting started, but it wouldn't be long before I started questioning whether this had been the right decision, and more importantly, questioning Cliff's motives.

CHAPTER 2

SPEND IT LIKE STANFORD

'Bigger than Branson'

The Redbus Group was created with a different mission to Cliff's previous businesses. It wasn't one brand with one focus and this time it wasn't about building a castle, it was about building a kingdom of data centres, starting in London and stretching across Europe. Cliff once claimed that he was going to be 'Bigger than Branson'[10] and made no secret of the fact that he wanted his business to imitate the Virgin Group. He chose the name Redbus because he said it was 'Easy to pronounce in any country and reflected the big red London busses that we all recognise.'[11]

So, like his rather thinner but equally bearded business hero, Branson, he diversified his interests across over 20 very different companies in a wide range of sectors, including Redbus Land-mine Disposal Systems (RLDS), a landmine clearing firm that, he claimed, could clear an area the size of a football pitch and make it safe in a day. He was so confident that he wanted Charlton Athletic (a Premier League football club at the time that Redbus sponsored) to play a match on it! He also created Serralux, a reflective glass manufacturing company and even a dental train-ing office! One of his better-known endeavours was the creation of Redbus Films, which made *Bend It Like Beckham* and *Maybe Baby*. At least that had a lasting degree of success, unlike his stab at being a pop mogul with failed pop band Girls@Play. Cliff had thrown £1m into making the band a success, but despite featuring Amstrad boss Sir Alan Sugar's niece – EastEnders star Rita Simmons – the band bombed after a short stint in the charts at number 19.

Such was Cliff's elaborate plan to take over the world with the Redbus Group, that he actually commissioned a London dou-ble-decker bus to be driven from London to the tip of South America to promote the brand. Funded by a range of sources, the trip was planned to cover 18 countries in eight months, ending in Tierra del Fuego, Argentina.[12] Of course, rather like many of Cliff's overzealous PR ideas, the trip never actually hap-pened.

However, despite his over-the-top marketing campaigns, there was little success across the majority of brands in the Redbus Group. So Cliff's attention moved to the one business that still had plenty of promise, and the chance of jumping on the dot-com gravy train; a small internet hosting company called Redbus Interhouse. This soon became Cliff and John's baby. They were still riding high on the success of their sale of Demon and thought that in time, they would be able to do the same with Redbus Interhouse.

I could see why they were so excited about the new business. Redbus Interhouse appeared to be the perfect business at first, listing on the London Stock Exchange in 2000, raising around £220m when it launched under the ticker symbol RBI.L.

They had used the money raised at IPO to expand the brand by building Redbus Interhouse data centres all around Europe – London, Paris, Frankfurt, Madrid and Amsterdam – with plans for global domination. However, what they hadn't predicted was that, as a consequence of the dot-com disaster, many businesses who would have previously needed data facilities were now going bust, and the expected wave of companies needing hosting capacity in data centres never materialised. So in 2001, Redbus Interhouse had virtually no customers and the stock price came crashing down.

Despite the obvious cause for concern, Cliff maintained his bullish confidence and either ignored the reality of the situation or simply didn't believe it was that bad. Instead, he was relishing the media attention he was getting from his promise to be 'Bigger than Branson'. He had even been boasting about his potential future success and comparing himself to the famous entrepreneur, as John Cassy reported in *The Guardian*. He said, *'I admire Branson tremendously, but I told him a while back that we're going to do in six years what it took him 20 to do. He's been content to muscle in on a mature market and grab a 10% share. I want 100% of new markets.'* [13]

I don't know what Branson made of all these comparisons, but I do know that Cliff believed his own hype, and his behaviour was fuelled by his own words. He did everything to excess; partying with celebs and the likes of Max Clifford and eating out at places like The Savoy. He even spoke with the familiar rasp of a man who'd smoked for more hours than he'd slept. His public profile had grown, and his ego was getting out of control (as I soon realised once I began working at Redbus).

If I were a rich man

Unfortunately for Cliff, his ability to make money was vastly superseded by his ability to spend it, and without anyone but his 'sleeping' partner to govern his behaviour, he was in danger of becoming a victim of his own success. When asked to justify his outrageous spending sprees and luxury lifestyle, he would tell anyone who would listen that he could do what he wanted. I suppose he was at least honest about his ongoing hunger for money, saying, 'I always wanted to be wealthy. Money doesn't buy happiness but lack of it can lead to misery. Now I really enjoy my life, I do whatever I like, and no one tells me otherwise.'[14]

It was no surprise to find out that the first thing Cliff did after he sold Demon was to move over to Brussels (presumably for tax efficiency purposes), and then buy himself a Rolls-Royce Silver Spur with a hefty price tag of £150,000. It wasn't long before he added to his indulgences with the regular use of a private jet and a new villa in Spain. It's not quite Virgin Airlines and Necker

Island, but whilst Cliff was challenging Branson to be crowned Britain's biggest entrepreneur, he was also challenging the bank balance at Redbus Interhouse by charging much of this to the company!

With no profits at Redbus Interhouse, the cash raised at IPO was certainly not going to sustain his spending behaviour. Later, in a shocking sequence of events, Cliff would also find that having multiple homes overseas might be an elegant way to save on some tax, but it does not alleviate a business leader of his financial responsibilities to his shareholders, employees and customers. In fact, Cliff's homes in Spain, Columbia and Belgium would later become central to an investigation into where the Redbus profits had actually been spent. Cliff's team of builders, who had been working on the 'substantial renovation'[15] of his Spanish villa, were allegedly paid for by the company shareholders.

Trading places

Now you might think, with all this going on, anyone would be mad to take a job at Redbus. But let me remind you, that at the end of 2001, when I was first considering the new role at Redbus Interhouse, I didn't have all this background knowledge about Cliff and John to base my decision on. I hadn't been back in the country long, and I wasn't aware of the press on either of them, so had no reason to deep dive into their personal histories, either on or offline. I put my faith in their word and trusted that they would want the same for the business as I did – to grow and expand sustainably, with a strong team on board and a robust plan to scale Redbus Interhouse to great heights.

So in February 2002, I innocently signed my new contract and became the Redbus Interhouse Sales and Marketing Director. Kevin Neal was the CEO at the time. He had been Cliff's boss in his former business and he now found himself with the shoe on the other foot, working for Cliff. As mentioned before, Kevin wasted no time in placing his three sons in the business. One of which was Simon Neal, who he made UK MD, reporting to me.

When I joined, Cliff and John were equal shareholders and the company was worth £6 million with a share price of just under 2p, down from over 300p at its peak. I needed to see what was happening financially before I could make any decisions on the sales and marketing strategy. To my horror, I discovered that Redbus Interhouse was actually worth less than it had in the bank, which was around £6.2m. The reason for that was even more shocking though, as it was burning £2.3 million per month in cash and fighting a losing battle to stay afloat, let alone recoup its investments.

It was clear that my job was to bring in the desperately needed sales that would plug the holes in the finances and deliver a turn-around in the business's fortunes. However, it became rapidly apparent that this needed to be done effectively in less than three months, or the company would be out of cash! I have always been someone who relishes a challenge, and I'd turned the fortunes of businesses around before, and I was comfortable I could do it again. But this was different. Even if I went out and became a complete overnight selling success, the company would not have been able to get enough cash into the business to save it in the three months we had left before we ran out of cash. I needed a plan that would encompass more than just sales. This required a completely new business strategy if we were to have any chance of succeeding. Or I needed a miracle.

Cliff, meanwhile, was undeterred by the rapidly decreasing cash flow, and buried his head firmly in the sand, maintaining his determination to grow the business, while *still* trying to prove to everyone that he was going to be 'Bigger than Branson'. However, his extravagant 'speculate to accumulate' mentality just depleted our cash flow further. It was like he was spending play money – building, buying and opening new Redbus data centres around Europe like they were hotels on a Monopoly board. On top of that, the construction industry has always been notorious for 'leaking cash' around builds and tenders. This was no exception.

In Cliff's eyes, he was building his portfolio and taking advantage of the growing need for small and medium-sized businesses to outsource their data systems cost-effectively.

In my eyes, he was letting his ego get in the way of any common sense, and the big red bus needed a new driver! John was also appearing increasingly uncomfortable with where things were heading – though as it later turned out, this was the least of his problems, he definitely had bigger fish to fry.

Champagne supernova

During the introductory phase of my new job, before my first day in the office but after I had signed my contract, Cliff asked me to come on a private jet tour of Europe to meet the candidates for all the country MD roles. These MDs were to be my direct reports, so it made sense that I should meet them and give my approval prior to their appointment. It was an experience that I would never forget and one that became par for the course on the Redbus journey. Spend. Spend. Spend.

The plane we travelled on was a 7-seater, it was me, Cliff, Kevin, Claudia Luque (Cliff's Columbian girlfriend who he'd hired as Head of Marketing) and a guy from Ogilvy (the renowned PR company Cliff had hired at great expense to raise our profile across Europe). There were bottles of Veuve Cliquot champagne rolling around the aisle of the plane, and Cliff and Claudia were getting it on with her sitting on his lap, totally undeterred by anyone else's presence. The whole team and the two pilots were smokers. I was the only one not smoking, but found myself stuck in a small smoke-filled metal tube unable to even open a window. It was more like a Rolling Stones after-party than a business trip. I sat, watching it all unfold, wondering what on earth I had let myself in for by taking this job.

We eventually landed and started the business side of the trip. Over the next few days, we visited all the European sites in Paris, Amsterdam, Frankfurt, Milan and Madrid, but despite the hype, I could see that what seemed like an exciting, success-fuelled growth strategy to the outside world was, in reality, a disaster waiting to happen.

Each day we were in a new capital, and I met the soon-to-be country managers who would report to me; Martin Essig, Adriaan Oosthoek, Luca Beltramino and Eric Lehoucq. In the evenings, we would dine in the finest restaurants, and in the morning we would wait in the hotel lobby for Cliff and Claudia to come down

We had just a few paydays left in the bank before the business would go bust

before jetting off to the next venue. They were always late, but that's okay because the jet would wait…

When we came back to the UK, I joined the company as planned and then started to take a good look from the inside… I laid all our cards on the table for Cliff and John to see. Our CFO was a guy called Carl Fry. He showed me the books. It showed that we had just a few paydays left in the bank before the business would go bust and explained that something needed to change. Now!

My colleagues remember this time as one in which Cliff and Kevin seemed to have absolutely no concerns about the money in the business – they acted as if another few million would just turn up out of nowhere at some point. The first time they were aware that the business was in danger of collapsing was when I told them! They had been heavily involved in the business, but here I was with a fresh outlook and an inquisitive mind, and I wasn't just going to accept what I was told – I wanted to see what was happening for myself and I wasn't afraid to share what I found!

After just a month in the job, I walked into the board meeting with an announcement to make. I was not a board director, but I had been asked for an update on the sales, and it wasn't good news. The board was shocked. They had clearly either misunderstood just how bad things were, or Cliff's endless optimism had been hard to see through. Either way, now they saw the facts, they had questions and they wanted answers.

At the time, I thought perhaps management had been telling the board that everything was on track. Perhaps Cliff was saying that he would provide further capital for the company if it was needed. But whatever the message was, the board had been hypnotised into believing that the business was in good shape, whereas nothing could be further from the truth! Spend was on budget, but revenues were a disaster and the company was burning cash at an alarming rate.

The dream-team partnership between Cliff and John was quickly turning sour as John woke up to the facts he'd obviously been avoiding. There was also now the question of whose fault it was to consider. The blame game had started. Who was responsible for ignoring the reality for so long? Heads came out of the sand and it was developing into a clash of the titans as Cliff and John disagreed with each other's decisions and both began to fight for ultimate control. With expenses rising, sales almost non-existent, and two leaders at each other's throats, the wheels of the big Redbus were about to come off in a spectacular way.

Costs needed to be slashed dramatically, but with such big egos to manage, getting both partners to agree to anything was going to be difficult at best, and impossible at worst. I made it clear to Cliff and John that the only way we would survive would be for them to stop bickering and step away from their executive duties to enable me and my team of executives to take the lead in the rescue operation.

But, before we could rescue anything, we needed to establish where the company's cash was being spent. It wasn't hard to guess who was spending it! We also needed to close as many operations as we could and a fresh plan for new capital investment to keep the company alive.

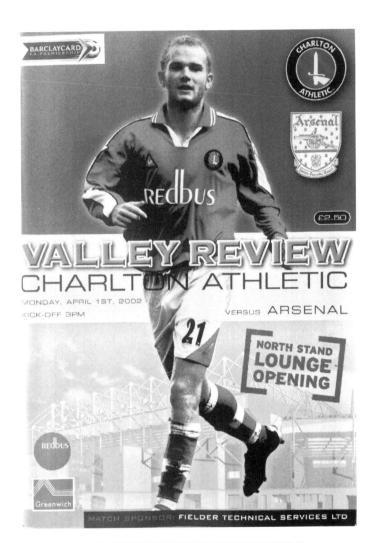

REDBUS-SPONSORED CHARLTON ATHLETIC MATCHDAY
PROGRAMME ON THE DAY WE FLEW IN THE FRENCH

Sadly, my financial warning had little impact on Cliff's attitude to spending. He saw and heard only what he wanted to, and carried on as before, ignoring my recommendations. On the 1st April 2002, he arranged to bring a group of journalists over from Paris to visit the Redbus Interhouse docklands facilities. It was an elaborate affair considering that we were trying to turn the business around and didn't have any extra cash to splash around.

First, we were flown by private jet to Le Bourget airport to meet the group. In true Cliff style, and trying to emulate Branson again, we put on quite a display with Redbus branding everywhere – even on the jet's tail. Together, we were flown back to arrive at London City Airport around 10am, and when we landed, we were picked up in Cliff's black Rolls-Royce, which also had the Redbus logo proudly emblazoned on it. My role was to present the new Redbus Interhouse strategy to the group and give them a tour of the facilities, making them feel confident in our abilities, before moving on to an afternoon of corporate entertainment at Charlton Athletic Football Club.

Redbus was Charlton's main shirt sponsor at the time, at a not-in-significant cost of £2.5m. Cliff compared the company to the Premiership side saying, 'Like Charlton Athletic, Redbus is enjoying a period of tremendous growth and looks forward to making a real contribution to Charlton Athletic as a Premier League side.'[16]

Our logo was all around the stadium and on the players' shirts, so after our meeting, we drove to the ground to impress them by watching Charlton versus Arsenal. Of course, being French, they all wanted to see Thierry Henry play. The car purposely dropped us 100 metres from the stadium so our guests could walk the last bit and see everyone going to the match in their Redbus shirts. One of them actually turned to me and asked if Redbus owned London. Cliff would have been so proud.

We watched the game from the directors' box and enjoyed a lavish few hours drinking champagne, eating and relaxing before flying our guests back home on the luxury jet. Arsenal won, and the French journalists saw Thierry Henry score a couple of goals so everyone was very happy. It was certainly a surreal experience, and you had to hand it to him, Cliff's ability to project a positive financial impression was outstanding, even if it was not actually affordable!

Dancing with the Devil

Back to reality, and I knew that what I called 'Operation Save Redbus' was not going to be easy. Alongside running the daily business operations, we were constantly managing the press, who were more than a little bit interested in the fortunes of the company and its two increasingly controversial founders.

Cliff had become a bit of a media magnet over the years, having a big personality, and an even bigger appetite for a lifestyle more akin to a rock star than your average internet start-up founder. Meanwhile, John's family had a situation that was bubbling away under the surface, which unbeknown to us, was soon to become a major contender for the attention of the media.

Cliff had recently been featured in the *News of the World* after being photographed leaving a London Restaurant with two table dancers. When the 'kiss and tell' story broke, the women

THE SUNDAY TIMES · JUNE 2, 2002

Election looming: John Porter, left, disagrees with Cliff Stanford over the future direction for Redbus Interhouse, a sponsor of Charlton Football Club

Redbus egos clash in boardroom

Nick Rosen

THE two founders of Redbus Interhouse, the internet hosting company, are locked in a boardroom row over the future direction of the company.

Deputy chairman Cliff Stanford, a millionaire whose exploits with a pair of table dancers once aroused newspaper interest outside the financial pages, wants to become chief executive.

But he is opposed by chairman John Porter, son of the former Westminster council leader Dame Shirley Porter, and the row now threatens to overshadow the annual shareholders' meeting on June 26.

On May 15, Stanford and Porter both relinquished executive duties and the new executives, including Michael Tobin, sales director, proposed implementing a cost-cutting regime to put Redbus back on its feet.

Stanford believes that cost-cutting is not the way to take the company forward and asked to be made chief executive at a recent board meeting so he could follow his own strategy.

At the meeting, Stanford, who netted £33m in 1998 by selling the internet service provider Demon Internet to Thus, the telecoms group, lost the election by six votes to one. He is now expected to use the company's annual meeting to try to hold another vote.

Porter said this weekend: "We have to focus on building shareholder value. That comes from having a strong platform and showing we know how to operate in a cash-positive manner."

Stanford wants Redbus, sponsor of Charlton Athletic Football Club, to stick to last year's spending plans and has posted his views on internet bulletin boards.

Since selling Demon, Stanford has become renowned for a lavish lifestyle. He was pictured in the News of the World stepping out with two table dancers and the report talked of "champagne-fuelled romps".

SUNDAY TIMES' COVERAGE OF THE SPAT BETWEEN PORTER AND STANFORD

described how they had been caught being naughty in the back of his car during a 'champagne-fuelled romp'[17]. *The Independent* also reported the story, but their angle was that rather than admit his misdemeanours, they said Cliff seemed to 'positively revel in the image of the Playboy'[18].

Cliff batted off the suggestion of any wrongdoing, 'It was a set-up,' he said. 'The Colombian girl was a friend of my ex-wife. We went out for dinner and were walking back. She got £40,000 from it apparently. Well, good luck to her. It didn't really do me any harm.' He went on to defend himself in his usual roguish style, saying, 'I've never had two hostesses in the back of any car. I've never even had one. Perhaps I am missing out!'[19] Considering the photographic evidence, Cliff's rebuttal of the accusations fell on deaf ears, but you couldn't help but find him oddly amusing, even if he attracted attention for all the wrong reasons.

'Nothing like a Dame'

As it transpired, the press had just as much interest in John Porter as they did in Cliff Stanford, only with a very different focus. Born into a wealthy family, John's mother, Dame Shirley Porter, was the daughter and heiress of Sir Jack Cohen, founder of Tesco supermarkets, and wife of Sir Leslie Porter. She was also the former Conservative Leader of Westminster City Council.

Dame Shirley Porter and her family had enjoyed the perks of their high profiles during the 1980s, thanks to her rise to political fame, resulting from her creation of local environmental campaigns. In 1990, she secured the title of Lord Mayor of Westminster. In 1991, she was appointed Dame Commander of the Order of the British Empire by John Major after delivering a spectacular victory in Westminster for the Conservatives in the 1990 elections, as reported by *The London Gazette*. In 1993, she retired from office and moved to live in Israel with her husband.

In the late 1980s, whilst she was leader of the council, Dame Shirley Porter created the Building Stable Communities policy, which focused on getting support from the eight marginal wards where the Conservatives wanted to get into power. This was developed after the Conservatives narrowly escaped defeat in the 1986 election. The policy was later deemed illegal, and the campaign became known as the 'Homes for Votes'[20] scandal, with much of Westminster's council housing sold off for commercial sale, instead of it being re-let to eligible families when it became vacant. As a result, many of the properties lay empty for months and even years and were kept secure from squatters at great cost to the taxpayer.

In 1996, after an investigation into the scandal, the Building Stable Communities policy was deemed illegal by the District Auditor and Dame Shirley Porter was accused of gerrymandering – which is 'manipulating the boundaries of an electoral constituency so as to favour one party or class'. She was accused of moving people who were less likely to vote Conservative, including the homeless, hostel residents, students and nurses, out of the wards that needed more Conservative votes and into Labour areas – giving developers their buildings to create private dwellings for wealthy professionals more likely to vote Conservative. All at the expense of the taxpayer.

The scandal was labelled the 'worst case of corruption in modern local government history'[21] in the press. And Nicholas Lezard described Dame Shirley Porter as 'The most corrupt British public figure in living memory, with the possible exception of Robert Maxwell.' in his review for *The Guardian* of *Nothing Like a Dame: The Scandals of Shirley Porter* (Granta, 2006), which was a powerful expose written by Andrew Hosken of the *Today* programme, whose own personal investigation documents Dame Shirley's incredible rise and fall.

Dame Shirley was ordered to repay taxpayers a total of £43 million. But she denied any wrongdoing and claimed that, despite her heritage and visible wealth, she had no access to the kind of money she would need to pay her fine, admitting only to having a few hundred thousand pounds worth of assets to her name. Of course, it seemed impossible that the heiress to the Tesco millions did not have more personal wealth than she was disclosing. A long and expensive investigation ensued, with Westminster Council battling for years to trace Porter's millions.

Over the course of the late 1990s, there were TV documentaries, books and numerous news pages dedicated to the case, with claims that Dame Shirley had moved all her assets into offshore accounts and overseas investments. At one point, the legal battle was costing Westminster Council so much money they felt it was no longer viable to continue the search. But in 2001, the House of Lords reinstated the case, with Porter's movements, communications, and financial dealings being scrutinised once again in the treasure hunt for her fortunes.

At the same time in 2002, we were trying to stabilise our ship, and although we all knew the story, we didn't think Dame Shirley's life or money had anything to do with Redbus. So, except for a bit of passing interest, we didn't focus on the news. We were busy fighting a multitude of fires, and it definitely wasn't a topic John wanted to have a chat about over coffee. It did, however, bring another noteworthy character into our world. Peter Green.

John had introduced Cliff to his financial adviser, Peter Green, back when they'd sold Demon for £33m – well before Redbus began. Green was an independent financial adviser to the Porter family and hence had responsibility for structuring Dame Shirley's investments. Cliff was naturally impressed by his promises to make good returns on the fortunes he'd made from Demon and started working with him to make his own investments. Green's advisory company set up trusts in the British Virgin Islands for both Cliff and John.

I didn't know it at the time, but with two such dramatic characters leading the business, and the influence of people like Green who operated in their circles, we seemed to be writing our very own pantomime. Looking back, it was no wonder that everything we did in the business attracted so much attention. The boardroom soap opera that unfurled over the coming months and years created more press headlines than any one of us could have ever predicted. The story was only just starting to unravel for me, and in the years to follow, every character would become instrumental in the future of Redbus Interhouse and the data centre industry itself.

CHAPTER 3

THE BATTLE FOR REDBUS INTERHOUSE

'Sex slurs fly on top deck of Redbus'

In Summer 2002, whilst we tried to rein in the business costs and get the finances under control, John and Cliff become locked in their own battle for control – of the whole company. They had previously agreed to my strategy of both taking a backseat whilst the management team got the business under control, but that only lasted a few months before Cliff felt that John interfered too much and was a 'disruptive influence' on us – so he decided to get himself back in the driving seat.

The problem was that although John agreed with my prudent stance to reduce costs in order to save the floundering company, Cliff was adamant that cost-cutting would not take the business

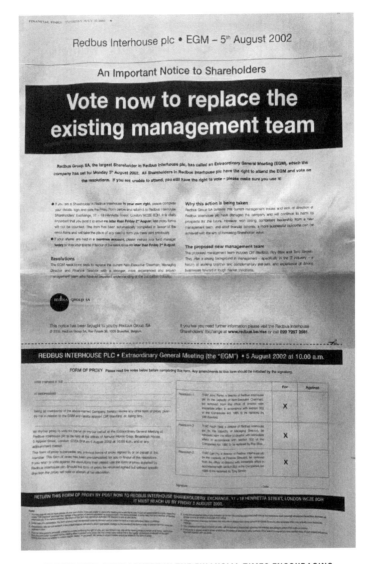

CLIFF'S FULL-PAGE ADVERT IN THE FINANCIAL TIMES ENCOURAGING EVERYONE TO KICK OUT THE MANAGEMENT TEAM

forward. Let's not forget though that Cliff was clearly not shy of spending money, and he had previously wasted millions of pounds on glamourous marketing ideas. He maintained his 'build it and they will come' attitude, as if money were no object. But tension had been building as they both had their own distinct ideas about the future direction of Redbus, and eventually, and inevitably, the volcano erupted.

Cliff stormed out of the building in a fit of rage, resigning his position as deputy chairman and director of the company, and Cliff's non-exec colleagues, Tony Simkin and Alex Bligh, also left the board. With Cliff now gone, his long-time partner in crime, CEO Kevin Neal, followed suit. The board was reformed, with John Porter as chairman, Carl Fry as FD and me as CEO.

Cliff's dramatic exit from the board was, in fact, just the beginning of the fun and games. He started contacting shareholders telling them that the dire predicament of the business was down to John and me, completely ignoring the disastrous period of waste that he had presided over which had left the company in such a mess.

But Cliff soon had a change of heart, deciding that he was better off in the company, and everyone else should be out of it, rather than the other way around. He wanted John out and for him to be made chairman so that he could revert to his own strategy of growth at all costs. He was probably missing his plaything too.

Unfortunately, as a 30% shareholder, Cliff was able to call an Extraordinary General Meeting (EGM), demanding a shareholders meeting to establish his position. An EGM is a meeting called to discuss special business measures or circumstances, so you would only expect to call one very rarely, if at all. Over the course of my time at Redbus, I think we had around seven EGMs! Cliff knew he wasn't the popular choice to lead the business, so he went to the unusual lengths of taking out a full-page advert in the *Financial Times* to try and rally support from shareholders!

Cliff lost the vote by a surprisingly narrow margin and John and I, together with the FD, Carl Fry, hung on. But Cliff was now as dangerous as a wounded beast.

Cliff's anger at losing the vote only fuelled his resolution to continue the battle for control. His determination to oust John quickly rose to the next level and Cliff started fighting dirty, trying to discredit us, his opponents, with a series of public slurs. It was an awful feeling being accused of things in the media. There was nothing I could do about it. If I commented on it, it just fuelled the flames... if I ignored it, people assumed the media stories were true!

I was living in a small village in Kent at the time, and one morning I opened my front door at 5am to the sight of a cameraman and interviewer asking for a comment. I also had a call in my car on the way to work from Radio 5's *Wake Up to Money* programme, asking me to respond to questions, and Cliff was in the studio live!

To help him, Cliff pulled in the services of publicist James Hipwell, a City guy who worked for Max Clifford, to use his expertise to spread dirt about John. The outrageous claim he came up with was that John was having a homosexual affair with Bo Bendtsen, our new non-executive Director.

Fortunately, Bo saw the funny side of the claim. When the story broke, he was in Hong Kong with his girlfriend. He told us that they 'fell on the floor laughing at the absurd claim'. I said to the press at the time that it was pretty ironic that Cliff would make such an accusation, considering that he was sleeping with at least one of our staff members at the time!

Bo was an interesting character. He was the founder of a business called UK2.net, which was an early customer of Redbus Interhouse in Docklands. He had a vested interest in the survival of the company as it provided mission-critical hosting to his business. Bo was an innovator and technical genius who would later go on to be a co-founder of JustEat.com. He was also Danish, and when I first met him I thought he had a bit of a personality bypass, but more of him later.

Cliff and I were far from seeing eye-to-eye by now. He was a massive distraction and timewaster, and what we really needed was to stay focused on the business before it went under. My strategy was to try and keep him at bay whilst we reduced employee numbers and shut sites to save the rapidly drowning company. It's never pleasant to make redundancies, but without making cuts we would have nothing left and we only had a few weeks left to get everything sorted if we were to have any chance of survival. At least this way we would be left with something.

So, with a damaged ego, and angry that despite all his bizarre efforts he was still not getting his way, Cliff turned his attention to me. To my horror, but not quite disbelief based on his history, he tried to muddy my name with a shocking sexual assault allegation.

Of course, when the accusations were made public, they were a complete fabrication, but this was a serious accusation and my reputation was at stake. I was accused of sexually assaulting a woman who worked for me, who herself had told me she had been having an affair with Cliff at the time (whether it was true or not, who knows?). I was put under investigation by the company. The plan she concocted was that she had set up a lunch with a client and asked me to attend with her. She would then drive me in her car back to the office where she would secretly alert her husband (with a predetermined signal) who would then contact the police saying that she was being attacked. They could find us because, for that week, she had rented a car with a tracking device, and I would be 'caught in the act', or at least that was the plan.

The plan, however, went awry when she got drunk at lunchtime and I refused to get in the car with her. She had lost the plot and sent the signal to her husband anyway and he dutifully alerted the police. Apparently, however, they subsequently swooped on the car only to find her in the drive-in McDonald's car park munching on a burger and arrested her for being drunk in charge of a vehicle. Well, we all get hungry after a drink!

Fortunately, when you work in a data centre you are surrounded by hundreds of cameras that are watching over every part of the building – and the footage proved that I was in a completely different place to where the claimant had said the assault took place. She later admitted to me on a call that Cliff had paid her to make the accusation. Finally, the investigation against me was dropped, and I never did find out for sure if Cliff was really behind it all. I just wanted to get on with my job now and stop all the drama, which was getting us far too much negative attention. I guess we will never know about that one…

A truly Extraordinary General Meeting

Cliff didn't give up and called a second EGM, this time for August 2002. In order to turn the 'bus' around, we had to stick to our vision to keep reducing outgoing costs and tightening our belts. And we knew the EGM would be a waste of time and another distraction from what we should have been doing to save the business, but we vowed to continue to stand up to Cliff's desperate attempts to unseat the board, and John was firmly on our side. Not least because in July 2002, Cliff surprised us by serving a High Court writ on the management team, claiming that John and other members of the board had breached the company's Articles of Association. In the writ, he also sought to block John from chairing the EGM.

In an interview with *The Independent* about the falling fortunes of Redbus Interhouse, Cliff said:

'We got the thing started off right, pointed people in the right direction and then stepped back. Unfortunately, the management has lost its way. They have been late with every site opening and they have failed to adapt to a changing market. Three years down the line and they've got no new products, no new pricing, and they've made financial mistakes that have cost the company millions of pounds[22].'

Redbus Interhouse
Letter to Shareholders

Redbus Interhouse PLC
30 July 2002

Under embargo until 0700hrs on the 30th July 2002

Redbus Interhouse plc

EXTRAORDINARY GENERAL MEETING

The company announces that the following letter has been sent to shareholders.

Michael Tobin, Bo Bendtsen and Paul Dumond have written:

'Dear Shareholder

Extraordinary General Meeting at 10.00 a.m. on 5 August 2002

'We are writing to you again because complaints and questions have been raised
by the company connected with Mr Stanford which has requisitioned the convening
of the EGM as to whether the letter sent to you on 19 July 2002 should have been
written by those directors whose removal Mr Stanford's company seeks, whether
the letter should have contained the unanimous recommendation that it did and as
to whether those directors should have participated in the deliberations of the
Board as to the convening of the EGM, the contents of the letter and associated
announcements to the stock exchange.

'We do not consider any of these complaints or questions to give rise to any
substantive concerns. Nevertheless, in order that there can be absolutely no
doubt as to the proper implementation of the process and that shareholders can
be in no doubt as to the recommendation, we write to you again. In order that
this letter of recommendation is a standalone document, we shall repeat the
matters which we consider to be relevant to your decision. None of the
directors to whom the resolutions relate has participated in the deliberations

YET ANOTHER EGM ATTEMPT. LETTER FROM ME AND DIRECTORS REFUTING CLIFFS CLAIMS... AGAIN!

Cliff was going to town trying to discredit us to both the general
public and our shareholders, and behind the scenes, he was
angling to bring his ex-Demon executives Roy Bliss and Tony
Simkin in to help him take control. He wanted Dan Wagner, the
former Dialog and Smartlogik entrepreneur to be our non-executive chairman, and to get Peter Randall on the board as a
non-exec too. We weren't too worried as we seriously doubted
that he would get the votes he needed to succeed, and our
legal team confirmed that we had not breached the Articles of
Association, so the meeting would go ahead under John's chairmanship. We prepared shareholders with a letter explaining our
current situation[23].

The press, meanwhile, were having a feast with all the stories, slurs and statements flying around. Remember, this was a public company listed on the main market of the London Stock Exchange. Perhaps what was most ironic though was the fact we were fighting over a company that had once been valued at £200m, but since the dot-com disaster had become worth only a few million at best. The share price was less than 2p per share, down over 99% from its listing. Basically, it was a junk stock. The company was actually worth less than the cash it had in the bank because it was burning it so fast! People were speculating that Cliff was desperate for money as he must have spent all the millions he'd made from the sale of Demon, and this was his last hope.

On the 5th August 2002, the EGM took place at the offices of Ashurst Morris Crisp, our legal advisors, in Appold Street. During the meeting, I explained that I had no choice but to trigger the strategic savings plan, reducing staff numbers from 300 to 80. I had people swearing and spitting at me when I had to fire 220 of our people. None of the other directors of the board had the balls to stand with me and announce it, but it was the only way to save the business. I also cancelled other unnecessary company spending – getting rid of the £2m marketing budget that had brought in a few measurable results and was frankly embarrassing considering our turnover was only £9m. The cuts included the sponsorship of Charlton Athletic FC, which we sold to All Sports, the loss of heads of department, including Claudia, Cliff's girlfriend, and many more jobs across head office, marketing and administration, saving £4.3m and delaying what was becoming an ever-more likely bankruptcy.

Cliff still seriously regretted his decision to step down and tried to take control back once more, but by now I had befriended John Porter and we were ready to defend against any further onslaught. There was talk of him now bidding to buy the whole business, but thankfully that fell quiet as he flew out of London to lick his wounds after the EGM. He decided to hatch his next plan in his luxurious Malaga home.

Sink or swim

Relieved that the EGM was over and exhausted by all the internal battles we had been fighting, we were now finally able to start making some business progress. We had £6 million in the bank, and the share price was around 1.75p. It was still a monumental uphill challenge just to get this business to survive, let alone thrive. I really had no idea how we were going to do it. What would my next move be? I had a few thoughts, but little did I know that an Act of God would get in there first.

It was just a week after the EGM, and whilst the atmosphere at our UK headquarters was brightening up, the commercial situation was still looking bleak. To top it all, one of the projects I couldn't mothball or stop was the Prague data centre. It was built and ready to open. We even had Czech Telecom (CT) due in on day one as a customer. But that was it, and we had offered CT six months rent-free as an incentive to attract more customers into the data centre.

There was no way we could make this data centre profitable, but I had to fly out to Prague because the British Ambassador, Anne Pringle had already agreed to do a big press launch. All I

**PRESS CONFERENCE AT THE OPENING OF THE REDBUS PRAGUE DATA CENTRE
WITH THE BRITISH AMBASSADOR, ANNE PRINGLE, AND THE
CZECH COUNTRY MANAGER, BORIS BELOUSOV**

could see was another £250k per month cash burn! We opened
the site to great fanfare and the local press were in attendance.
I remember getting on the plane back to London and thinking 'this is bumpy' as we took off just as a storm was brewing.
Central Europe was about to be hit with the worst floods it had
seen in decades.

Now known as the 'Hundred Years Water', the natural disaster
struck over a couple of days; claiming lives and destroying homes
and businesses across the region. Soldiers were deployed to
erect flood barriers, thousands of people were evacuated, and
even the city's zoo animals were taken away to the safety of dry
land. Sadly, nothing could stop the torrent of water travelling
through the city. Twenty-nine of Prague's metro stations disappeared underwater, all road and rail links were cut, footbridges

CITY DIARY

city.diary@thetimes.co.uk

REDBUS Interhouse, the internet hotel group, seems remarkably sanguine given that its brand new facility in Prague has been forced to shut down due to flooding. Although this means that some of its customers' websites are in danger of going down, the group insists that it is not a big issue because all its customers and most of the website users are also under water. "We're all in the same boat," says Mike Tobin, Redbus's sales and marketing director, without a trace of irony.

THE TIMES' ANNOUNCEMENT ON THE FLOODING OF THE PRAGUE DATA CENTRE were destroyed, and buildings collapsed. The devastation was unbelievable.

Despite all our cost-cutting, we were still contractually obligated to open and operate our Prague site, thanks to our one and only (free) client, Czech Telecom! The site had only opened a couple of days before the flood, and with the opening ceremony still fresh in my mind, I got the call from my brand-new Czech country MD, John Polak, that the floods had breached the front entrance of the data centre and continued to rise at a pace!

I remember my first call with him. He was in the basement. That was where all the high-voltage transformers and diesel generators were. He told me he had evacuated the building but wanted to check the HV equipment had been turned off by the National Grid operator. He called me and said, 'The water's coming through the bricks! What should I do?' I said, 'Where are you?', and he said, 'I'm next to the switches, holding some cardboard over them to keep them dry'. I said, 'My God! Get out!'

After switching off the power and evacuating his whole team, he made his way up to the offices and grabbed a few personal items. The water was already flowing through the lower stairwells and was now two metres high in the main foyer of the building. Eventually, he got to the roof, where he spent the next 36 hours with only a dying phone and some Kendal Mint Cake for company – refusing to abandon ship before he was rescued by the fire brigade boat. Now that's dedication!

In the end, the floods quite literally washed our Prague operations out. The water level peaked at seven metres in the main foyer of the building. We suffered a huge amount of flood damage and were unable to provide any services to our customers. But, as I always say, everything happens for a reason, and although traumatic, this was also serendipitous.

Part of our rescue plan for the business had been to pull back on growth and consolidate our sites – and despite only just opening, Prague was one of them. We'd had to go ahead with the launch of the site thanks to legal obligations, but we knew we would need to find a way to close it quickly to help consolidate costs. But we needed cash, so this unexpected Act of God literally saved the company. I used the opportunity to turn our fortunes around and approached the insurance company with a deal that would save us all from going under. I said to them, 'Give me £8 million now and I'll go away.' It was a bold move to save the business but considering that a new data centre costs around £50 million to build and ours had just been destroyed, we could have insisted on a much larger pay-out, but we would have been up against a long legal battle to get that level of money for flood damage, as well as being in a long queue of people claiming for the same issues. We just didn't have time. We needed enough cash now to keep us afloat before we sunk for good, and the insurance company were happy to give it, knowing they had got a pretty good deal too!

At last, we finally had the cash we needed to keep the business going until the next round of investment came through. Knowing what we now know about how this event changed the course of our future forever, you could say that rain is the single biggest reason why Redbus became so successful!

Keeping our heads above water

After the floods and the chaos of the summer, finally, thanks to the insurance pay-out, we had enough money in the bank to cover our operating costs, which allowed us to gain some degree of control and stability. We got through the winter by maintaining a freeze on spending and selling off other sites, including our Luxembourg facility – which was bought by a company who wanted to use it to store physical papers and data! Ironic for a data centre, but it was a big site in a small country, so it made sense for us to sell and for them to buy.

We'd also had to extract ourselves from a legacy contract that had been signed with the leaseholder of our proposed Munich data centre years before. We'd been left with an empty warehouse that we were contractually obligated to fit out, and which demanded rent of £1.5 million per year, for the next 15 years!

Martin Essig, our German country manager based in Frankfurt, travelled to Munich with Carl Fry to meet the owner and try and strike a deal. He remembers the trip well:

'We needed to tell the landlord the dire situation Redbus was in, but on the weekend prior to us going Carl had put his back out. He was in extreme pain but knew we had to get some kind of exit plan on this unit, so we went anyway and the air around him was just death. There was no way Redbus could survive if we had to pay this lease.

We walked out of there with an agreement to pay them just a trivial amount of money. I think we paid one year's rent and gave the property back in a state that wasn't particularly

"

The very thing that was our problem, our over-specified top of the range infrastructure, was now our differentiator!

good because all the car park area had been ripped up to put in power cables. Probably the amount of money that they got from us for that lease was just enough to put the building back in the state it was when we had taken over the property! So, that was a really, really important day for the entire group, and an important day for Carl that he was able to get us out of that contract.'

Turnover was also on the up, and by the end of 2002, we had increased revenues by 24%. The previous overspending on top-dollar equipment had become a sunken cost. The internet was starting to become important to certain customers and the fact that we had awesome equipment meant that we were their preferred supplier. The very thing that was our problem, our over-specified top of the range infrastructure, was now our differentiator!

Most businesses paid by direct debit, and once they were in, they were unlikely to leave. They were connected to dozens of other customers in the building. Almost impossible to move. When the telecoms operator KPNQwest ceased trading, it announced to its customers that their data centre would be switched off, so we contacted their customers to migrate their equipment over to us. But even then, almost all of them waited until the power went off before moving over.

We had large-scale global customers like AOL and BT by this time, so business resilience was key – combining disaster recovery and business continuity to get them through any unexpected situations, without stopping business or losing customers. Attitudes had changed, and businesses had realised that storing their mission-critical data in a data centre was the way forward, especially since 9/11, when the need for a disaster recovery process became very apparent. Our customers could rent an EMC multi-terabyte back-up box, basically, this is a massive data storage box, which meant that if there was a sudden disaster and everyone had to work from home, they could all access their mission-critical information, giving them a 'belt-and-braces' solution.

When the company was IPO'd it was done via a reverse into a cash shell called Horace Small Apparel, which used to make workwear like uniforms and overalls in a factory in Nashville Tennessee. At the time, this was a unique event and allowed the company to effectively buy a non-operating business that had cash in it, with the proceeds of our IPO. It didn't matter that the company had nothing to do with data centres. We inherited the disused and abandoned factory, but couldn't sell it because it was the subject of an Environmental Protection Agency investigation. About 100 years before, the land that the factory was sitting on had been used as a railway siding where steam engines were parked. Over the years, the oil had seeped into the ground and when it rained would come to the surface. So, we couldn't sell the asset until we cleared the land. This gave me yet another problem for the business to overcome, with regular trips to Nashville to progress the case. I became familiar with all the bars on Printer's Alley, and aside from a few memorable beers with a great local lawyer called Brooks Smith of Boult Cummings Lawyers, it was just a bunch of headaches for not much progress!

But here we were, we'd made it to 2003. We were keeping our heads above the water and still surviving – just. The insurance money had kept us alive for another five months or so and during that time we were gaining customers and cutting costs, but we were far from self-sufficient. We needed new investors, and fast.

CHAPTER 4

EXORCISING THE DEMON

All change!

By March 2003, I had formally been announced as the CEO of Redbus Interhouse, and the likes of Kevin and Cliff were consigned to the past... or so we naively thought.

We were actively looking for investment in order to take the company private, and successfully gained three new investors over the year; a Russian, an American and a Dane, who would all prove important to our success.

Bo Bendtsen had joined us back in 2002 as a non-exec director (shortly before Cliff accused him of having a homosexual relationship with John Porter – thankfully, that didn't put him off working with us). His company UK2.Net was the UK's largest web hosting company, with around 435,000 customers. Bo had developed a good relationship with John Porter, who was also a UK2.Net director. They were also linked through John's investment vehicle i-Spire[24], through cross shareholdings etc. i-Spire was invested in a bunch of different things, including both the Dance Portal for the Ministry of Sound and Cheapflights.com. In 2002, they were suffering their own challenges with a fall in their share price from 80p to 3.25p and the resignation of their auditors, PWC, due to 'irregularities'[25].

For a time, while he was growing UK2.Net, we had let Bo run his business out of a room in our Docklands office at a good rate, and he used to ask for my help with recruiting staff because his people skills were not quite up there with his techie skills! Over time, he had got to know us at Redbus, and he had been keen to invest, primarily to ensure the survival of our company as it ultimately kept his main business UK2.Net up and running! So in 2002, he became part of the 'rescue consortium' that would ultimately take us out of the financial toilet bowl.

And he was a great addition to the board as he was hugely bright technically. I remember he always wore black jeans and a black T-shirt. For over 18 months, we never saw him in anything else. Eventually, we said he needed to brighten up his act. Now he wears colour.

By 2003, John Porter had taken a bit of a back seat with the business and came into the office less and less frequently. He was a keen golfer and I used to see him cleaning his golfing shoes in the office occasionally. As his post was being delivered to the Redbus office, I started to collect his correspondence, but binned it after a while as he never came in to collect it.

One day I saw a members' diary come in from Queenwood Golf Club. Queenwood is known for its A-list membership, with people like Hugh Grant and Michael Douglas gracing its greens. Joining fees are reputed to be around £200,000 plus a yearly membership fee, so it's a great place to look for life's high-rollers!

Ironically, Queenwood is also known as one of the most secretive clubs in the UK, with people needing a password to even look at their website! Imagine my surprise when a member's welcome package arrived addressed to the CEO of Redbus, which naturally I opened. It contained the members' list for Queenwood with everyone's home address, email and phone numbers on it!

Passing over the phone numbers of the Gillette family, Michael Douglas, Catherine Zeta-Jones and so on, sure enough, there was John Porter. I had a look at a few more names and there was one I recognised from a conversation I had overheard John have recently, with a guy called Oliver Grace Jr, who was based in New York.

I had nothing to lose at this stage, and just needed people ready and willing to invest to hear me out, so I called the number for Oliver Grace. When he picked up the phone, I slightly elaborated on the truth, saying that John Porter had asked me to get in touch with an interesting business opportunity. Thankfully, Oliver didn't hang up on me, he wanted to know more and flew over from America to meet us. Kerching! We had another investor on board.

Oliver then brought in his friend Boris Jordan, an American businessman with Russian ancestry who had just been fired from being one of Putin's close aides.

Boris had a highly successful but somewhat controversial background. He was at Credit Suisse until he founded the Renaissance Capital investment bank in 1995 – Russia's first Western-style investment bank. When it opened, he'd held a lavish party with the US stunt water-skiing team flown over for entertainment. When he left, he took his core team with him, including a super-smart guy called Sergei Riabtsov, and created his own investment vehicle, Sputnik.

Such was Boris's influence on Russian business, he had also been a director at the Moscow TV channel NTV at a time when there was a horrific hostage situation in a Moscow theatre. On 23rd October 2002, around 850 people were held hostage in the attack by Chechens, demanding the withdrawal of Russian forces from Chechnya. Special forces were deployed and the stand-off lasted three days, during which time at least 170 hostages were killed. Eventually, a chemical agent was pumped into the theatre by the Russians, killing many insurgents and hostages. NTV filmed the crisis, and being half-Russian and half-American, Boris was accused of filming it to show Russia in a bad light.

Clearly, Boris had weathered a few storms. But he had also worked with Oliver in deploying fibre infrastructure around Moscow, so he was keen on tech, and specifically, internet-based investments. He saw a good future for Redbus, so Sputnik became an investor and Jordan took his place on the board.

ME MEETING MIKHAIL GORBACHEV IN NEW YORK IN 2003

Boris was a great collector of Russian art, and he once invited me to the launch of an art exhibition showing his collection in the National Gallery in New York. It was a rare display of artefacts and documents that traced centuries of Russia's evolution, and he had funded the exhibition to help Americans and Russians foster better relations. There was a celebration dinner held, at which I found myself sitting next to Mikhail Gorbachev and Rudy Giuliani, the former Mayor of New York.

Never one to shy away from controversy or miss the opportunity to capitalize on a transformative movement, nowadays Boris has moved on and focuses on his $1billion stake in Curaleaf Holdings Inc., the biggest cannabis company in the US. Today, Boris is described by *Forbes* magazine as 'The Cannabis King' and the world's only 'Pot Billionaire'[26].

Here we go again!

In March 2003, we announced our intention to delist from the main market and move the stock to the smaller AIM market. We wanted to do this because the market cap had become so small and the almost non-existent liquidity in the stock meant there were no institutions able to invest in us. AIM was focused on much smaller market cap companies and had a cheaper operating model than the main market. Then, in June 2003, Cliff appeared again, with yet another EGM. Another attempt to oust the board – in particular, John Porter. What he didn't realise, however, was that John had actually resigned more than two months earlier. We all had a bit of a giggle at that, but on a more serious note, we knew that Cliff was just stirring up more trouble. He had bowled into the meeting confident that he had the backing of the majority of shareholders, including institutions like Schroders, but for the third time running he was disappointed. When he didn't get his way, he immediately threw his toys out of his pram and called another EGM, which was to convene in three weeks. As a result, we also had to postpone the move to AIM.

On the 10th July 2003, we prepared once more to do battle with Cliff. And after an exhausting year of infighting, tensions were high. This was the seventh EGM in just over 36 months! It was again held at the London offices of Ashurst Morris Crisp and attended by around 40 people, including the board, shareholders and investors.

We had to block the entrance to the meeting to keep non-shareholders out, as we were aware that the press were desperate to get the inside scoop, not just on the result, but on the people

Redbus Interhouse
Statement re request for EGM

Redbus Interhouse PLC
02 June 2003

For immediate release, Monday, 2nd June 2003

REDBUS INTERHOUSE PLC
('the Company')
EGM

The Company announces that it has received a request from Redbus Group S.A. for an Extraordinary General Meeting. Redbus Group S.A. are proposing the removal of Michael Tobin, Carl Fry, Paul Dumond, Bo Bendtsen and Oliver Grace as directors of the Company, and for the appointment of Cliff Stanford, Tony Simkin and Roy Bliss as replacement directors. The request for an Extraordinary General Meeting also called for the removal of John Porter as a director, although he is not in fact a director of the Company.

As shareholders are aware, Redbus Group S.A. proposed similar resolutions in Summer 2002 following repeated attempts by Cliff Stanford to be appointed Chief Executive. These proposals were rejected by a clear majority of shareholders who voted.

Enquires:

Redbus Interhouse plc   020 7001 0000
Michael Tobin
Carl Fry

Weber Shandwick Square Mile   020 7067 0000
Terry Garrett
Josh Royston
sqmile@interhouse.redbus.com

This information is provided by RNS
The company news service from the London Stock Exchange

HERE WE GO AGAIN! THE EGM THIS TIME TO AGAIN OUST ME AS CEO AND THE ENTIRE BOARD OF DIRECTORS, AND REPLACE THEM ALL WITH CLIFF AND CO! IT ALSO ASKED FOR THE REMOVAL OF JOHN PORTER, EVEN THOUGH HE WASN'T EVEN A DIRECTOR BY THIS TIME!

in the room. All the time we had been fighting fires at Redbus, interest in Dame Shirley Porter's hidden fortunes had been growing, and John Porter was now under considerable public scrutiny. Cliff was always great media fodder too, so reporting on these two characters was top of the agenda for the paparazzi. Our boardroom battles had become pretty notorious by now too, so who wouldn't want to get into this meeting and watch the fireworks?!

The meeting lasted an exhausting five hours. There were many heated exchanges, Cliff even tried to get a vote of no confidence in Paul Dumond, our non-executive director, who was chairing the meeting – prompting one of his own supporters to walk out in disgust. It felt like a movie. It was surreal. There were people shouting from the back of the room. You could hear the demonstrators outside. They had unfurled banners around Ashurst's offices calling for 'Justice against Dame Shirley'. People were swearing at each other and shoving each other.

At one point, it felt like the meeting had been hijacked by supporters of Westminster City Council. One member was even removed for waving a banner that said, 'Pay up Porter'. We also discovered a journalist who'd managed to get into the meeting by buying a single share! He was removed when he started taking photographs that he didn't have permission to take. He was manhandled and bundled into the toilets. The film was removed from his camera to stop him publishing the pictures. He did eventually leave with everything he'd come with though, including the now overexposed film and a good story for the newspapers, albeit with no pictures to show for his time. Thankfully, digital cameras didn't exist at that time!

Finally, the EGM came to an end. After trying to sack the board and install his own people, again, every single one of Cliff's eight resolutions was defeated. The existing board won with 67 million votes, whilst Cliff got 61 million votes. At last, Cliff had to accept that he was never going to be able to get rid of us, and after three years it was time to stop trying. But Cliff being Cliff, he wasn't about to relinquish his shares, or stop there.

The sad truth is that if it wasn't for Cliff, we would have had a pretty simple, and increasingly healthy business. But by now we felt like we were just constantly dealing with a spat between him and John, which had escalated beyond control. It was extremely draining. I remember a journalist coming up to me at the end and asking me for a final comment. I said, 'I hope he now goes and enjoys his life in Malaga, sitting in the sun, and lets us get on with building some value for him'. It became quote of the week in *The Daily Telegraph* newspaper on the 11th July 2003[27].

Being the tenacious man that he is, Cliff's next move was to launch an investigation into various shareholders who had not voted for him, to try and find an issue that would change the result. Little did he know that the hunter was soon to become the hunted.

With the EGM over, summer 2003 should have been a calmer time for us. We wanted to focus on showing the shareholders that they had made the right decision to keep the current board, that we were still growing in value and doing everything possible to make sure profits were heading in the right direction.

But we knew that so long as Cliff was still there as a shareholder, our lives would never be easy. I was in Madrid looking over the accounts of the Spanish subsidiary and noticed an invoice for delivery of diesel fuel to Malaga. Odd... as our diesel engines are located in Madrid! But of course, Cliff had a villa in Malaga.

We were suspicious about some of the high expenditure that had been happening over the previous years, so we started an investigation. It didn't take long for us to find out that Cliff Stanford had not only been funding his lavish lifestyle with more than a little help from the business, he had also made £85,000 worth of renovations to his Costa del Sol home that had been generously paid for out of the company bank account. We took steps to reclaim the money, suing Cliff for the full amount.

Battleships

Enough was enough. The non-exec directors and I met on Boris's yacht off the coast of Long Island and came up with a plan to get Cliff off our backs and extinguish his ability to cause the company pain once and for all. The offensive had begun.

Meanwhile, Cliff's legal fees fighting the company were starting to mount up and eventually, in September 2003, he sold his remaining shares for around £3m – at a significant loss, to... Boris and Oliver.

But you just couldn't keep Cliff out of the news. This time though, it was through no fault of his own. In 1999, Cliff's Spanish au pair, Rocio Wanninkhof, was brutally murdered whilst walking home from a fair in Spain. Her body was found three weeks after she disappeared, and Cliff offered a large reward[28] for information leading to the conviction of her murderer. The events that followed were described as one of the biggest miscarriages of justice in Spanish history.

Rocio's mother's lover, Maria Dolores Vazquez, was originally convicted of her murder and sentenced to 15 years in prison. However, there was no actual evidence to convict Maria, and a retrial was ordered when DNA found on another young girl murdered a month later was matched to DNA at the site of Rocio's body. Cliff doubled his reward to the equivalent of £100,000 as the hunt for the murderer began again. Eventually, in 2003, a Brit named Tony King, otherwise known as the Holloway Strangler, was convicted of the horrific murders and sentenced to 36 years in prison[29]. As far as I am aware, Cliff never had to pay up the reward money.

And yet there was even more drama to come. What we didn't know at this stage was that this wouldn't be the last major court case for Cliff in the years to follow.

Crying Wolff

Ever since the earlier EGM on the 5th August 2002, John had been increasingly suspicious that his emails were being hacked. He shared these suspicions with his mother, but she clearly did not have the same level of concerns about her communications, as unbeknown to us, over the past year, she had sent increasing numbers of emails to John, in which she discussed their personal investments.

In January 2003, our own suspicions were raised as to how Cliff always seemed to know what was going on in board meetings he had obviously not attended, and how information on the performance of Redbus was getting out onto public forums without any announcements from us. We started to think we either had an inside leak, or Cliff was up to something.

We contacted the Police National High-Tech Crime Unit, who coincidentally were located in the same building as us, on the top floor. They agreed to look into our concerns. Over the course of a long investigation, and a later court case, the details came out in full.

It transpired that due to the high level of publicity our EGM had attracted in 2002, Cliff had been contacted by an ex-Met Policeman, George Liddell, who'd been following the story and claimed to be a 'Corporate trouble-shooter who sorted out boardroom battles[30]'. Liddell said that he could help Cliff remove John Porter from the Redbus Interhouse board, and to reinstate his own chairmanship, all for a tidy no-win-no-fee of £150,000.

Cliff thought this was the answer to all his problems, and bit Liddell's hand off to get started. He intended to expose John and his mother's hidden assets to the press, disgracing John to shareholders, and forcing him to finally give up his seat. This would allow Cliff to be reinstated on the board, and then he would subsequently get rid of me and the rest of the board.

To start the process, Cliff accessed all John's emails through an old computer server, sending two years' worth of correspon-dence over to Liddell on a CD. Liddell then instructed him to divert all emails coming in and out of Redbus to his own secret email account. Cliff did exactly that, and from the 10th Septem-ber 2002, Liddell received copies of all of our correspondence, as well as private emails going back and forth between Dame Shirley and her son, John.

We had been using a relatively obscure backend email system that Cliff himself had implemented in the early years, and he had secretly put a loophole into the system. It worked like this: he logged into the mail server as us, either John Porter, Carl Fry or me, and then had all the emails mirrored to an external account before logging out again. So, when we looked at any activity on the accounts nothing untoward was showing up. It simply showed logins from the three of us into our own accounts. At the time, we were all logging in and out of our email accounts all the time, and as Cliff was logging in with our login details, nothing untoward came up.

Liddell then started phase two of the plan – a smear campaign. Calling himself 'Wolff', he posted information he found in our emails onto online forums. There was one in particular called 'ii' or the 'interactive investor[31]' bulletin board, which focused on AIM stocks. He would post board decisions, financial information and all sorts of confidential stuff. Sometimes, information would appear on the bulletin boards before our board meeting had even happened because he would manage to get hold of the agenda in advance! Our head of IT, Paul Bruff, had the boardroom professionally swept for bugs and, of course, found nothing.

However, after he started doing this he panicked. He realised that if we ever found out, we could track the movements back to his computer. Alarm bells had started ringing and he told Cliff that it all had to stop. He was right, we were onto him. We'd discovered the diverts set up on the email, changed all the passwords and reported the crime. I couldn't believe this was happening! I had escalated the situation and I was now meeting with the National High-Tech Crime Unit of the Metropolitan Police more often than I was meeting with customers. It was now just a matter of time…

During the course of the investigation, our lead detective, Detective Geoff Donson, would sit outside relevant people's homes and see who met with who. It's what's referred to as a stake-out. One person they discovered Cliff had been meeting was Peter Green, his financial advisor. This wouldn't have been unusual, except that Peter wasn't just working for Cliff, he was also John's financial adviser. And, as we know, the Porter family were under great scrutiny by Westminster Council who were on the hunt to prove that Dame Shirley had hidden her assets, in order to force her to pay her £47m fine.

Peter obviously knew exactly what was going on with John's investments and, more importantly, where Dame Shirley's money had gone. Cliff had worked out that Peter Green had all the information he needed to be able to expose and shame the Porter family, but couldn't get Peter onside to tell him what he wanted to know (his work with the Porters was worth far more to him in the long run), so Cliff tried to pressurise Peter to tell him what he needed to know instead. Cliff was, after all, a man used to being able to pay for whatever he wanted in life, and assumed this would be no different.

We immediately realised what was happening and spoke with Peter. We told him we knew exactly what Cliff was doing. Peter's loyalty was clearly with John, as realistically most of his business was with the Porters, so he agreed to meet Cliff again, but this time wearing a wire. I wasn't told of this at first. It was done between the police and Peter. I was told after the meeting, and this came as a big shock to me. I thought this only happened in spy movies!

The meeting went ahead, and we heard it all, word for word. Cliff intended to force John out of Redbus Interhouse by exposing what he had found in the diverted emails to the press. We had it on record. Yet the verbatim recording still wasn't enough to convict Cliff unless we had the testimony of Peter Green to confirm the fact that it was a recording of his meeting with Cliff. This was just one piece of the puzzle, and to make sure this vital evidence was considered in court we knew we would need Peter to agree to appear as a witness to testify that this was him on the wire, and the circumstances surrounding the meeting.

Cliff's intentions had been to get any dirt he could on John, but in doing so he had unwittingly unravelled some of the mystery of Dame Shirley Porter's hidden wealth. This was a bigger result than even he had hoped for and made him feel somewhat above the law.

Events were about to take an even darker turn...

Hacked off

Going back to early 2003, with over 70 emails from Dame Shirley to John in their possession, Cliff and George had been getting impatient for a result. They decided to make their move and took the emails to Stephenson Harwood, the investigators working on behalf of Westminster Council, hoping to demonstrate that they were doing a 'public service' for the good of all by revealing what they knew (notwithstanding the illegal means of obtaining the facts!). In their minds, the next step would be for John Porter to be investigated – putting pressure on him to step down from Redbus. Simple!

But their plan backfired. John Fordham at Stephenson Harwood was suspicious of how they had obtained the information, and

Their cover was starting to slip, and the hackers were feeling the pressure

of their motives, and didn't take them seriously. They needed to move to Plan B.

Their cover was starting to slip, and the hackers were feeling the pressure. The conspiracy that Cliff had created with George had grown difficult to manage, and their impatience was about to put them all in the spotlight: Peter for his connection to the movement of the Porters' money, George for his part in the email interception and blackmail plan, and Cliff for initiating and agreeing to fund the entire project!

Cliff then made the bold move of telling Peter exactly what he knew about John and Dame Shirley's money. This was his final attempt to blackmail him to force John to step down from Redbus to avoid any public exposure for both himself and his mother of their financial dealings, and to give control of the company back to Cliff.

Again, it didn't work. Neither Green nor the Porters were prepared to be blackmailed. They had weathered too many storms to be pushed over so easily. They knew that the unlawful way in which their emails had been obtained would be Cliff's downfall, and Dame Shirley had spent a lot of time and effort staying one step ahead of her pursuers, so she wasn't about to hand herself over because of an overzealous blackmail attempt by her son's obnoxious business associate! The email that perhaps revealed most about Dame Shirley Porter was one about a surprise party for John. She told him: 'You appeared to be totally surprised unless, of course, you've inherited my genes and know how to lie?'[32]

Busted

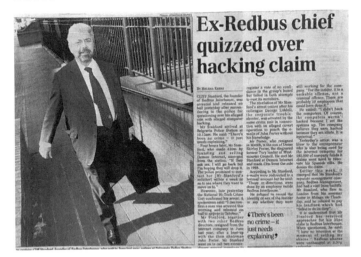

Ex-Redbus chief quizzed over hacking claim

GOT HIM! ON 12TH FEBRUARY 2004, CLIFF WAS FORMALLY CHARGED WITH 'CONSPIRACY TO BLACKMAIL AND INTERCEPTION OF COMMUNICATIONS UNDER THE REGULATION OF INVESTIGATORY POWERS ACT 2000'

On the 18th March, evidence had been building and Cliff and George's impatience for a result (so that Cliff could force John off the Redbus board, and George could get paid) had aroused enough suspicion and attention for the investigation to close in around them. Blissfully unaware of what was about to happen, George Liddell arranged to meet Peter Green in the Grosvenor Hotel, still hoping to strike a deal that would force John out of Redbus. After a short conversation over coffee, Peter picked up his paper and waved it at two men sitting nearby. They were plainclothes policemen. George's plan had backfired and he was promptly arrested for his part in the blackmail conspiracy. With

George protesting his innocence, it would take a little while longer to prove Cliff's involvement.

Over the following months, the BBC started digging around for information on Dame Shirley's investments. And in June, Peter Green contacted us to say his computers had been stolen from his office in Mount Street, which caused a lot of speculation and disbelief. Then, on the 30th June, the *Today* programme aired a short documentary uncovering what they had found about Dame Shirley's hidden investments. The game was up.

On the 5th September 2003, Cliff was arrested at Belgravia police station under the Computer Misuse Act on suspicion of hacking emails at Redbus.

The investigation went on over the winter, and finally, on 12th February 2004, Cliff was formally charged with 'Conspiracy to blackmail and interception of communications under the *Regulation of Investigatory Powers Act 2000*'. George Liddell was also charged, and both men were bailed to appear in court in March[33]. It was the culmination of a long fight. Strangely, I felt sorry for Cliff in a way. Sure, he had disrupted so many lives with his lies and personal attacks, and he had caused me so much stress and almost bankrupted the company, yet he could be a very charming character and his undoubted vision was evident in the creation of the company in the first place. But, ultimately, I was very relieved to see this drama drawing to a close.

Cliff reverted to form, managing his PR in the same way he always had, by denying any wrongdoing. Helena Keers reported in *The Daily Telegraph* that he said:

'I don't want to say I'm a saint, but I don't need to blackmail anyone. If my getting those emails and handing them to Westminster Council is the problem, I would do it again. I think that's a victory[34].'

We all prepared for the court date and then, on the actual day, I received a phone call from Cliff. He was at the airport in Spain. I said, 'Cliff, we shouldn't even be speaking, why are you calling me? You should be here.' He told me that he had bent over in the airport and his trousers had ripped from front to back, and there was no way he was appearing at the Old Bailey with his backside hanging out! Cliff was a big man, and although frustrated, I couldn't help but laugh at his response. And so, the date was postponed.

While we waited to bring Cliff and George to justice, yet another twist of fate changed proceedings. In September 2004, Peter Green was found dead in his home having suffered a heart attack. He was 62.

A few days before his death, Peter had told me that he had been 'pushed down the stairs' at Green Park Underground Station and had hit his head badly. Earlier, he had told me that both his office and his house had been broken into, yet only the computers had been taken. He was convinced that it was connected to the case and that these actions were being taken by someone who was trying to get information first, and then ultimately trying to get rid of him. We will never know whether he was right or not, but the evidence that had been extracted from the wire he had worn became inadmissible in court without him present. This was a massive blow to us. Obviously, it was a terrible shock to hear that Peter had died, and alongside our sadness for his family, we were acutely aware that we had also lost our key witness.

On 13th September 2015, Cliff Stanford and George Liddell pleaded guilty to one count of 'Unlawful and unauthorised interception of electronic communications' at Southwark Crown Court[35]. The blackmail charges had to be dropped due to the lack of evidence and, in particular, that of Peter Green, who had died just days before. Both men received a six-month prison sentence suspended for two years. Cliff was also fined £20,000 plus costs.

Judge Geoffrey Rivlin, QC, said during sentencing that Stanford had arranged to intercept emails:

> 'In the hope that by gaining knowledge of the contact of John Porter and Shirley Porter that might be disadvantageous to them so that you could use that information in furthering your ambitions in relation to Redbus'.

Adding that:

> 'It is essential people, in whatever walks of life, and, of course, those running important businesses, should know that the integrity of their confidential communication should be respected, and... they will be protected from being hacked into by outsiders.[36]'

Cliff tried to appeal the conviction the following year, but his request was denied. The man certainly had balls, but he just never knew when to stop. The evidence was piled against him, but he was an accomplished liar and certainly wasn't going to let the truth get in the way of his big plans! As far as I know, Cliff was still aggressively active in 2014 with a Welsh-based 'ambulance chasing' call centre called Ifonic Claims Ltd, which has since been struck off, but the consumer action bulletin boards, which he so enjoyed using against us, were very active against him and that business[37]!

Meanwhile, Dame Shirley Porter came out of her self-imposed exile in Israel, and now back in the UK, negotiated a full and final settlement of £12.3m to be paid back to Westminster Council, saving herself over £35m in unpaid dues.

It had taken more than two years, but finally, the whole sorry story was over. I was hugely relieved, but the events had taken their toll on me over the years. Fighting Cliff was arduous, painful and exhausting, but it was an experience I would never forget and could only learn from, and it showed me that I had the strength and resilience to get through anything.

One great thing to come out of the debacle was that during the course of the enquiry, we built up a great relationship with Geoff Donson at the National High-Tech Crime Unit. Once the investigation was over, I asked him if he would consider becoming our Head of Security (goodness knows we needed one). He agreed and a year later, once he had completed his 25 years' service for the force, he joined us. He also gave me a framed piece of paper that the police had found when they raided George Liddell's house. It had the full 'war plan' drawn up and is a lasting reminder of what we went through and won.

Finally, it was time to concentrate fully on Redbus Interhouse and to take the business on a brand-new journey.

PART 2

THE WONDER YEARS:

Getting serious and going public

REBUILDING REDBUS

Holding our breath

There is no doubt that the people left working in the data centre game after the boom and bust of 1998-2002 were carrying great scars, and I was one of them. But these scars gave me and my comrades the advantage over newcomers to the market. Those of us who had survived the wreckage knew the risks, knew the pitfalls, and most importantly, knew how to overcome them.

If you were still a business leader in the data centre industry after 2002, you had two choices: either believe in the industry, dig deep, keep the faith and have enough vision to see it through, or get out. I was a believer.

"

Redbus, Telecity, Interxion and Telehouse were basically where the internet lived in Europe

Redbus Interhouse had suffered many blows over its short life-time, partly due to the dot-com crash, and partly thanks to Cliff's behaviour. But throughout every battle, we'd kept a strong hold on the business, and we were slowly but surely recovering from the financial impact of the previous few years.

By the start of 2004, Redbus, Telecity, Interxion and Telehouse were basically where the internet lived in Europe. Over 90% of the UK's internet traffic passed through our buildings in Dock-lands, with clients including ISPs, telcos, web hosters and major content providers, such as the BBC and Sony. Our key drivers for growth were increasing annual recurring turnover, whilst making the most out of vacant floor space in our data centres.

By mid-2004, our order book had increased, our share price was on the up, and our losses were reducing. We were Europe's most advanced independent colocation company, which meant we had the best data centres set-up for housing and operat-ing the servers and technology services our customers needed to support their businesses to run efficiently. We were a criti-cal facility for offering them versatile, secure and cost-effective facilities for their data and telecoms applications. And we were leading the market because we gave our customers the freedom of choice to use whichever telecoms, Internet Service Provider and Content Provider they wished. We had coined the term 'carrier-neutral'.

We had state-of-the-art facilities and our engineering and support were second to none, which meant we were pulling in big contracts from the likes of AOL, AT&T, Barclays and a whole host of big names. Cash was still king, but business was strong enough to get us cash-flow positive in every country. It was just a question of being able to hold our breath long enough to see off the competition.

California Dreaming

Let's not forget, this was the decade of the dot-com darlings, the era where tech companies grew fast, and tech leader's egos grew even faster. Dot-com companies had created a stack of paper wealth, and the people leading them had more money than they knew what to do with. All that cash that had become king in earlier years made people like Cliff feel like they were King Kong, and he wasn't the only one rampaging around the City causing all kinds of chaos. Cliff was just one of many tech leaders in this genre. Yes, his spending was wild, and his nights out were wilder, but to be fair, he was nothing in comparison to the big US tech stars who had started the whole dot-com movement in the first place.

If you check out the biggest tech scandals of all time, you'll find they are all set in the decade we were growing up in at Redbus[38]. From accounting scams to spying, strippers to sexual harassment, money-laundering to murder, it was all happening in the tech industry from 2000-2010, the decade of shame.

Coming in at a rather (dis)respectable number 10 on the top 10 tech scandals is one of the leaders we knew well over the years; Robert McCormick, former CEO of Savvis Communications. Robert's story later became legend as he was reportedly named 'The Lap Dunce' by the *New York Daily News*[39]. As the story goes, Robert was celebrating with some business associates in the Manhattan strip club 'Scores' – where he unwittingly ran up a bill for $241,000 on his American Express corporate expenses card.

Robert denied spending more than $20,000 (still not a meagre amount) and both he and Savvis refused to pay the Amex bill. The whole debacle went to court as American Express sued Savvis, Scores and McCormick, until a settlement was eventually reached[40].

Needless to say, Robert was swiftly moved on from Savvis, but his legend lives on to this day, and if you want to bring it to life you can see the inspiration in the film *Hustlers*[41], which tells the story of how Scores strippers drugged wealthy stock traders and CEOs before running up their credit cards.

So, perhaps without even realising it, these were the guys we aspired to be like. Just like younger siblings looking up to their big brothers and sisters, we were the new kids on the block and we were looking up to the big boys as our role models. Their glamourous, high-rolling, fast-paced, Californian lifestyles had become iconic in our industry – and if it was good enough for Silicon Valley, it was good enough for London!

Dot-com darlings

Now we were one of the dot-com darlings in Europe, and I was highly aware of the need to get some positive PR that would show we were taking business seriously and help us get the interest and trust of investors. We'd hit the ground running in 2004 and I made it my job to very consciously and visibly lead Redbus out of the darkness by shining a light on our successes, rather than allowing the press to keep dragging up the dank and dirty stories we'd had to withstand over the years.

Over the next few years, we were recognised with a number of awards, including Turnaround of the Year at the 2005 London Business Awards, and I was listed in various publications as one of the UK's leading entrepreneurs for my efforts.

Finally, Redbus was starting to attract attention for all the *right* reasons. We were proving our worth as one of the darlings of the new economy – showing the world that we could 'do dot-com' just as well as the US – maybe even better, and attracting big players to our growing internet hotels.

The Double 99 Club

Our closest rival by 2004 was Telecity, who were 45% owned by the British private equity group 3i. By now, we were both listed on the AIM market, having moved there around the same time from the main market, and we both wanted to go private. Following closely behind us were Interxion and IXEurope, but at the time, neither were as big as Telecity or Redbus.

Telecity was the brainchild of Mike Kelly, a Manchester University academic, who founded the company with colleague Anish Kapoor in 1998. Mike doesn't get the credit he deserves for what he did in the early years (in the late nineties). He really had the vision to see how important data centres were going to be in the future.

Based at the Manchester Science Park, Telecity was floated on the techMARK Index in June 2000, initially valued at £550m, but when demand shrank across the industry because of the dot-com disaster, their share price went into free-fall.

In 2003, Josh Joshi, Telecity's Financial Director, was brought in[42], when they had just three months left to survive. They were in a very similar situation to Redbus at the time. He now calls this the era of *'The Double 99 Club'*, explaining that 'Telecity was one of a very elite club of businesses whose stock price went down by 99%... and then went down by 99% again!'

Redbus was also in *The Double 99 Club* because we'd originally floated at hundreds of millions of pounds during the boom, and then dropped back down to around £6m in 2002.

Over the past few years, Telecity and Redbus had hovered around a similar valuation, slowly climbing in worth every year. But we were competing for the same business, and the demand, whilst growing, still wasn't enough to have multiple providers of capacity in the market.

Both companies now had data centres in London's Docklands, had more sites in Paris, Amsterdam, Frankfurt, and were key nodes of LINX (the London Internet Exchange), which is the network that connects over 80 countries around the world, allowing for increased routing control and improved performance. We couldn't have been more similar, and we all knew that consolidation would be the best way forward but agreeing to a partnership would prove impossible.

A match made in heaven?

We'd had talks with Telecity over the years about a merger, but they'd always broken down before an agreement could be made. The problem was that we wanted to buy them, and they wanted to buy us. We were very similar data centre companies, competing for the same business, so it made sense to consolidate and merge, but neither of us wanted to be controlled by the other. We were a good match as together we would become the 'game-over' market leader in all the main markets across Europe. Even in the secondary markets, where there was less demand, Redbus was present where Telecity wasn't and vice versa. It really was a perfect match! Yet we were struggling to find a way forward for the deal. Part of the problem was the strength of the personalities involved, and the folklore about Redbus and us, its leaders, that existed in the industry, which blurred people's vision when it came to making a deal happen.

At this time, I was CEO of Redbus, and we had strong primary investors with Boris, Bo and Oliver on the board. Meanwhile, the Telecity Non-Executive Chairman was Michael Hepher, previously Group MD at BT, and their CEO was Ricky Hudson. We were all somewhat infamous in our own right.

Telecity and their investors had seen all the publicity around our boardroom battles, and apparently, according to Josh, I'd developed a reputation as a 'Master of the Dark Arts' because I did business differently to the traditional City types. I was referred to as a maverick because of my leadership style, which was very alien to the more formal 3i team. The 3i guys didn't know how to take me, so they just didn't trust me. I would go along to their offices to negotiate a deal that would give us a positive result, and it would all just end up as bluff and double bluff.

The first time I went to meet 3i and Telecity, I got off the train in Manchester, walked along to their office, knocked on the door and sat down. Ian Nolan from 3i and Rupert Robson from Telecity's advisors Torch Partners were in the room. I started to say, 'Thank you for letting me come here…' and they jumped in with 'Before you say anything, we want to buy you!' Then all the hair went flying as we battled it out for who was worth more than who.

Each time we tried to come to an agreement, we'd get a bit further along the line and start doing the due diligence on each other, but someone would suddenly throw their hands in the air and say they just didn't like the other guys, and we'd be back to square one again!

If I'm honest, neither side wanted to give away the business just at the point the industry was coming to the fore, and the premium that one side would have to pay over the public share price to make the deal acceptable to the other would be too high. So, just as we'd get closer, one of us would try and scuff up the situation so that we could get the other cheaper. Josh said we were like 'street fighters' – some of them were more stealth-like, but I had nothing to hide and they could see me coming a mile away.

Deal or no deal

The Telecity team were trying to take the company private so they could use it as a vehicle to start to consolidate the data centre space, and we were doing the same thing. They had tried going to IXEurope for a deal, but they were too expensive. The banks wouldn't help either of us, and we were both public companies, so every time we so much as breathed, we had to make an announcement. It was all a bit of a mess, but it still pointed to one necessary outcome. Somehow we had to bite the bullet and make the merger happen.

In 2005, we'd spent a good year or so at loggerheads with Telecity, but by now it was blindingly obvious to everyone that the businesses needed to merge. And we were both finally at a cash positive position. But neither shareholder faction wanted to be a 49% holder of the combined business!

I got friendly with Rupert Robson and met him one night for a drink in a bar in St James's. He was working as an advisor to the other side, so it had to be done in a very clandestine way. If either set of shareholders had found out we were meeting to hatch a plan, we would have both been fired. I said, 'How the hell are we going to do this Rupert? 3i don't want to have a minority share. Redbus shareholders don't want to have a minority share, and a 50-50 split will be a disaster if the two sides can't come to an agreement on strategy. So this deal isn't going to get done!'

We came up with the idea of bringing in a middle agent to have a sort of ⅓, ⅓, ⅓ structure. Then any shareholder that did have a good idea could just convince the new entity, who didn't have any historic baggage or pre-existing interest in any particular

"

Telecity was conservative, process-driven, methodical. Redbus was cavalier, flexible, fast-moving

"

outcome other than to create value, and thereby have an operating majority for decisions.

Shortly after our meeting, Rupert came up with Oak Hill Private Equity as the answer to our problems. They were a US West Coast-based middle-market private equity firm. He put his idea to Telecity and Josh Joshi and Ricky Hudson met with Oak Hill first, explaining their frustrations and how they wanted to develop the market and achieve growth. Josh identified that we had different visions. Redbus was focused on gathering Internet Service Providers and connecting them to the internet. Whilst Telecity were focused on trying to provide corporate customers with an internet solution. Josh explains:

> *'It was a slightly different kind of customer base we were pitching for, and those backbone guys at Redbus were looking for month-to-month contracts. They were investing hundreds of millions of dollars in network infrastructure and architecture; and they wanted flexible contracts.*
>
> *Telecity were focused on signing three and five-year deals, building long-term business relationships with corporate clients, introducing annual pricing escalators, that sort of thing. That was much more akin to the kind of value we wanted to create with those customers.'*

As a result, Telecity, with Oak Hill and 3i on side, was delisted and taken private in September 2005. We knew that Oak Hill couldn't do two simultaneous PLC take-privates so it was logical to get one done and then add the second. We knew the plan and what the next steps would be...

The beauty parade

Redbus was going through the same take-private process with our own shareholders, and as our discussions with 3i and Oak Hill flourished, we eventually delisted Redbus in January 2006 and merged the two companies together to create the rather unimaginatively named TelecityRedbus. We were finally now a great business and we were ready to take on the giants.

We had also eliminated the board issues we'd had with Cliff, and we had world-renowned investors Boris Jordan, Oliver Grace, Bo Bendtsen, 3i and Oak Hill on board. And now we were free from the shackles of being a public company. But Telecity and Redbus were culturally very different businesses. Telecity was conservative, process-driven, methodical. Redbus was cavalier, flexible, fast-moving. This made the more conventional shareholders of Telecity nervous.

Josh Joshi and Ricky Hudson had obvious reservations about working with us, and, at the beginning, put up some resistance. However, they had been around the houses and knew that this merger was the right thing to do. Now we just needed to work out who would run the new combined company, and so began the beauty parade.

3i wanted their management team to run the combined business. They didn't like or trust me and the former Redbus team, and their conservative style and approach matched that of the former Telecity management. Boris, Oliver and Bo (let's refer to them as BOB for short) on the other hand, wanted their team, which was me and Carl Fry, my CFO. After all, we had all gone through hell together. Neither side wanted to back down, so we decided both management teams would pitch their vision for the combined business and the investors would decide which team to back.

It went like this; Ricky and Josh would present their strategy, then we would present our strategy – and we would each pick holes in the other's. We were fighting to demonstrate why each of us would be the best option to go with because both sets of investors wanted the guys they knew.

This kind of competition is pretty hard not to take personally. You had CEO going against CEO to see who the best leader would be. CFO going against CFO to see who would be better for sorting out the financials etc. etc.

I was up against their CEO Ricky Hudson, and our CFO, Carl Fry, was up against theirs, Josh Joshi. We were all at opposite ends of the spectrum personality-wise and it was difficult to know which way the investors would swing.

Our nickname for Carl was the Grim Reaper as he was a risk-averse, conservative kind of guy. He was the yin to my yang; my glass was always half-full, whilst his was always half-empty. You know the type. But maybe that's what made us a good combination, we must have complemented each other's strengths and weaknesses. But at this stage, no matter who they chose for each role, there would be a life-changing impact all round.

Fight or flight

The first pitches were so different. We had an upbeat, optimistic vision of the future. On the Telecity side, Ricky was dour and conservative. We clearly won. No competition. But 3i, and in particular Ian Nolan, wouldn't accept our vision. Ricky knew his pitch hadn't gone well, but said that 'his team wasn't ready'. What rubbish. Both teams knew what the stakes of this poker game were from the very beginning! Ricky had blown it!

Anyway, such was BOB's confidence in me and Carl, they agreed to a re-run a week or so later. What happened next was a bizarre twist of fate that would push one of the key contenders in the parade over the edge.

The away day

Rick Hudson decided that his team needed a bit of space and inspiration to build on their strategy, so he planned an away day for the Telecity management team and invited Oak Hill along.

They went to a beautiful hotel outside Oxford and were joined by J. Taylor Crandall, one of Oak Hill's 'Big Gun' founders (who is a multi-millionaire in his own right) and a number of others. They had flown over from the US in a private jet, landing just outside Oxford. It was an impressive entrance by all accounts.

What happened next was to be a pivotal moment in Telecity FD Josh Joshi's career. He tells how the story unfolded in his own words:

The notebook

'It had been a great day, we had a whole day meeting, a nice lunch and dinner, and we had a really good view on the strategy. The morning after we had a closing session and J. Taylor Crandall says, "Right, let's go... I'm heading back across the pond back to California. Anybody want to ride on my bus?" meaning "Do you want to come on my private jet?"

Of course, all the Oak Hill guys go, "Yeah!"

So, he says, "Well, I'll meet you all in 10 minutes. My car's waiting. You want to come, you're at the door in 10 minutes, I'll give you a lift on my bus to California."

So, they all charged out of the room. Bless them. I totally understand it. The UK guys scarpered off home as well.

I wasn't able to go to California, so I'm the guy at the end, cleaning up after the meeting. But when everyone walked out the room, I found a notebook, and realise it must belong to one of the guys who worked for Bob Morse. Bob was the lead on this opportunity from Oak Hill's perspective.

It was a classic black and red notebook, A4 style. I think, "Whose is that?" and I pick it up.

I put the notebook in my bag and boarded the train back to London. As I was sitting there, pondering on my professional integrity and how I had never put a finger wrong in my career, I also thought about all the shit I'd been through. I suddenly felt very naive.

I wasn't mature back then. I thought if you worked hard at something, it would eventually pay off. I genuinely *believed that the investors would back our management team, and that Ricky and I were the best choice for CEO and CFO in the merger.*

I couldn't help myself. I opened the book.

Then I swear to God, this point is imprinted on my brain and will never leave me. I turned the page and there's minutes of a meeting with Bob and Ian Nolan. They're having this meeting and they're saying,

"Right, so what exit package are we going to give Ricky?" – *my Chief Executive.*

"So we're going to do a deal. Mike will be... We don't think we can get this done. Mike will have to be the guy that will be the CEO. Ricky will..."

This is a meeting that they've had previous to today, our Oxford away day.

"We'll give Ricky a million pounds, but do you think a million pounds will be enough? [Ian to find out from Ricky whether a million pounds]"

Nothing about me, nothing about the CFO.

There were more comments. Then it was like this moment of going, "Fuck me. These guys are doing the dirty on us!" because I thought that they would actually back their management team. If they're getting rid of Ricky, does that mean they're going to get rid of me too? Really?

I felt totally betrayed. I thought, I'm screwed.

So, I got off at Paddington. Before getting on the Tube, I was withered with rage. I picked up the phone to Ricky and I said, "Ricky, I just picked up this notebook. I don't know what to do, but we're fucked. They're going to fuck you over."

And Ricky says, "Yeah, I know."

And I say, "You know? Why haven't you told me?"

He says, "Why should I tell you?"

It blew my mind. All I could think was, "I thought you and I were partners in this?" It turned out that I was just the help.

A week later, we had the second beauty parade and I knew the Oak Hill guys would be at the meeting. I didn't know what to do. I had spent so many years turning Telecity around and had been through confrontation after confrontation in my career. I couldn't face any more confrontations.

So I found the guy who had left the notebook behind and I looked him in the eye and said, "I think this is yours."

The colour drained completely from his face and he took it and walked away without saying a word.

I wanted them to reach out to me. To tell me they knew what I knew, and to explain. But they never did.

That burned me even more. We were working 18-hour days. It was incredibly hard work. I felt as though we were losing so I was being a bit more emotional, a bit more immature about it.

It was a two-day meeting, and that had happened on the morning of the first day.

On the evening of the last day of the beauty parade, it was clear that we were on the back foot. The Redbus strategy was winning. Bob took me out for a drink.

Bob just said, "Josh, I can see that you're troubled, but you tell me what you want. If you want to be the CFO of this, then we'll make you CFO."

I regard that comment to their credit, I think that they had realised that the strategy that they wanted to go for was Mike's strategy, and they thought that the CFO they wanted on their team, i.e. the person that would look after the books for them, would be me. I think that's how they saw it.

I was too emotionally gone at that point in time and felt betrayed by them. I said. "I can't do this." I really didn't mind much that it would be perceived I had lost to Carl. While I didn't agree with Mike's strategy, I wished him well and in the end, his vision created billions in value for Telecity shareholders. The bit that burned the most was that I felt betrayed by my own side. I couldn't tolerate that and pulled the ripcord – probably the best career decision I ever made... other than joining Dave Ruberg at Interxion.

The deal was announced, and it was confirmed that Mike and Carl were going to take over. We had a vehicle arranged to take Redbus private, and then to merge the two businesses, Telecity and Redbus, a new business would be formed, which was still to be decided. Mike and Carl were to be the CEO and CFO of the combined business.

I stayed for six months to be the lead management team member on the Telecity side to help with the integration. Ricky walked on day one. That really caught in my craw.

I had non-competes named in my contract, so I couldn't go to any of our competitors. So I went and became a CFO of an online gaming, gambling and sports betting business.

A year later, I'd had enough and I wanted to get back to the industry that had broken my heart and start fixing it.

I got a call from David Ruberg offering me a role at Interxion. I said, "Mate, I turn 40 in 2007. I'm going to spend my time with my family until the day I turn 40. But if you're ready to wait for me until then, then let's have a conversation."

And the rest, as they say, is history.

Ricky has since passed away, and I respected him dearly. Ricky was the man that turned Telecity around. It was his strategy that allowed us to actually succeed. I had my part in it, take credit where credit's due, but Ricky led that team to turn Telecity around.'

And that was the end of Josh's story at Telecity.

I had my own foresight of what was going to happen in the second round of the beauty contest. I was having an affair with one of the 3i partners' PAs, so I gleaned my information from her. I'm not proud of it now, and I wasn't as cynical as to plan it that way; it just happened as I was spending a lot of time at 3i's offices at Waterloo before they moved to Buckingham Gate in London.

The second round of the beauty contest was held in their main meeting room on the top floor. It was the last ever meeting there and we were told not to bother using whiteboards but just draw on the walls as the whole room would be stripped out the next day!

Needless to say, the second round of presentations went the same way as the first and Ricky finally had to admit defeat.

On the 3rd November 2005, the announcement was made that Redbus Interhouse and Telecity would merge to create Europe's largest data centre operator.

The senior management team selected to lead the newly merged businesses was announced as: Mike Tobin, CEO (Redbus Inter-house), Carl Fry, Finance Director (Redbus Interhouse), Trevor Wadcock, Operations Director (Telecity) and Matthew Gingell, Marketing Director (Telecity). Rick Hudson CEO, Telecity, and Josh Joshi, Finance Director, Telecity, will step down.

To this day, I believe that both sets of shareholders wanted me as CEO and Josh as CFO, but Josh had been caught up in the politics of it all. I would have been very happy with that team!

BUILDING A NEW TEAM

A new dawn

January 2006 heralded a new beginning with a newly combined team forming the backbone of the Telecity Group. We were starting to get used to each other and generally trying to put any friction and tensions to one side so that we could focus on the growth of the business.

Our joint forces made us stronger and our reputation was building. We now hosted the London Internet Exchange (LINX), the Amsterdam Internet Exchange (AM-SIX), and the Deutsche Internet Exchange (DE-CIX), the three largest exchanges in the world. And we had 18 data centres across Europe, covering 500,000 sq. ft of fitted space.

Our corner of London's Docklands was quite literally the most connected place on planet Earth. We even had our own avenue of stars along Marsh Wall, made up from the manhole covers of every telecoms provider from BT to Virgin Media, Colt Telecom, TalkTalk and Metronet to name a few. We hosted the content for Universal Pictures, Sony PlayStation's servers and the BBC iPlayer, and up on our roof were aerials and satellite dishes for the likes of Vodafone, CNN and Al-Jazeera.

We operated a lean structure with no excess costs, and our business model was simple and predictable, we only had four main costs: rent, electricity (which was passed onto customers), maintenance, and staff – of which we only had around 400 across Europe. I didn't even share a PA with anyone, never mind have one for myself. Even just a few months into the financial year, I could tell you what our full year's figures would be to the nearest £500k because data centres had a very simple income and expenditure model. The cost base is pretty fixed, and your customer base is highly recurring with very little customer churn.

There were often blurred lines between operations and board roles because we were all willing to dive in and do whatever was needed to do during the transition and restructure. But we did need to recruit for a variety of important roles, and after what we'd been through over the years, we only wanted the right people on board.

Testing the limits

My team from Redbus knew what I wanted and expected from them, and they knew the kind of personal values and strength of character I valued. The business had put them through their paces over the years, but I'd always made sure that I'd had a positive impact on the way they felt about themselves, their roles, and how they would manage the potential fear of a life-changing event such as our merger.

In our earlier years at Redbus, when I was building the team, I wanted to create an experience that would convert their fearfulness into fearlessness, by giving them the opportunity to do something that would instil fear in them, but then show them that their fear was, in fact, unfounded.

I decided that experience was to swim with real sharks – after all, they would soon be doing the same metaphorically, and they needed to know how to cope! By asking them to do it in real life, I was forcing them to confront their fears. They were thrown in at the deep end – quite literally.

Together, 13 of us travelled to Edinburgh, but only I knew what was in store for my team. After a strategy meeting and tour of a local whisky distillery, I took them to Deep Sea World where they discovered my plan.

After getting kitted up and having a short training exercise, it was time to get into the water. I didn't want to force anyone to participate, but I did want to see them overcome their anxiety if they could. Three out of the thirteen of us said they were too scared and that they couldn't do it. I didn't try to force them, I left them to make their own minds up. While they thought about it, the others were being taken two-by-two into the water by the instructors. There was a five-minute swim down to the seabed, just long enough to get really frightened. Then you saw the sharks.

These were big beasts. Three-metre-long hammerheads and tiger sharks. When it was my turn, the anticipation on the way down was definitely unpleasant, but once I was down there, I felt much better. It was extraordinary to be this close to the sharks with no cage or protection between us at all. They swam right up to me, out of curiosity, but well-fed by their handlers in advance, they were not looking for a little lunch.

We were advised not to touch them, and I hoped everyone would remember that because there was no insurance and I didn't want to lose anyone from the management team before I absolutely had to!

As the groups of two went down, those left waiting their turn had an exquisitely painful wait. They could see each pair coming back out of the water, some absolutely buzzing with the wow factor, others just glad to be out of there, though still exhilarated. In the end, everyone swam, bar one.

Once the shark dive was finished, we all went out to dinner. I asked everyone what they had felt when they realised what was in store. One guy said he had been excited, that it was something he had always wanted to do. The others said they had felt a huge panic. Everyone admitted to having been terrified – regardless of the fact that the instructors diving with them had assured them it would be fine. Nothing the shark specialists said had alleviated a primaeval fear.

And what about when they were on the seabed? That was still terrifying, they said, but nowhere near as terrifying as the anticipation and apprehension. As the seconds and then minutes ticked by and the sharks were only staring, not attacking, they had realised that what the instructors said was true. They started adjusting their expectations. When I asked how they felt when they came out, they told me it was 'life-changing' and one person added that they had dealt with something they didn't even know was an issue.

The one colleague who had not swum with the sharks said he now deeply regretted it. And more than that, he felt a sense of alienation because he could not share the experience with the others.

In the event, the Redbus management team had only lost one person in the merger with Telecity: and it happened to be that same guy. By his reaction to the challenge set, I think it was perhaps symptomatic of a general attitude.

Extreme team building

Another memorable trip I took the team on was to Iceland. Martin Essig remembers it well:

'We took two private jets over to Iceland from London, but we were delayed a couple of hours with bad weather. We had an appointment to have a Scotch whisky tasting at the top of a glacier as soon as we landed. The glacier was, according to Mike, a one-hour drive from the airport. It turned out to be a four-hour drive, and when we got there it must've been six or seven o'clock in the evening. It was pitch black.

Not one of us had ever done anything like this before. We were given crampons to put on our boots, and nobody really knew how to put them on. There were two guys to help us, but they didn't seem particularly knowledgeable either. Everybody got one of these single LED lights to put on their head. We walked up this glacier and everybody was still in their street clothes. I'll never forget that Brian McArthur-Muscroft, our new CFO, was in his tweed hiking jacket and cap. It was minus 10 degrees!

It must have been a 45-minute walk up. It wasn't terribly technically challenging as the mountaineering people would say, but it was for us. There were these holes. They looked like the shape of a funnel, and we were pulled away by the guides who said, "Don't go near those, they go down to the base of the glacier, 100 or 200 metres below, and if you fall down them, there's just no chance that you're going to get out!"

There weren't many of the holes, they were just peppered around. I'd say every 30 or 40 metres there'd be one of these things. We had a few shots of whisky at the top and we were walking back down, and we're all tired and freezing cold with these stupid useless lights on our heads. I was walking down with Brad Petzer our Financial Controller and at some point we just kind of stopped and looked to our right. We were literally a metre away from one of these holes. We looked at each other and said, "This is without a doubt the most dangerous thing we've ever done in our lives."

This was what Mike called "Team Building". If it hadn't worked, it would have been team shrinking! We always did these things. I'm very grateful for it, having been able to experience a lot of things that I would've certainly never done in my life otherwise.

So after these kinds of experiences, when we were given the brief by Mike to run our data centres (and this is something that I know that my colleagues and I all agree on) we knew this opportunity was probably never going to happen to us in our careers again if we didn't take it now. We were new competitors with really established players in markets and a dramatically shrinking customer base. Not expanding but shrinking. All of these dot-com companies and telecommunications carriers were disappearing at the time. But we all just sort of adapted our own businesses, and we had the opportunity and the trust of Mike and the shareholders to just do whatever we could to try to salvage something after the dot-com boom and bust period.

We were able to shape our businesses to how we thought that they might work in our individual markets. After four or five years of doing that, we ended up having very, very distinct businesses in each of the different European countries. All of them very successful. Niche players, with very high margins, especially compared to our competitors.

That was something that was hugely empowering for all of us, but we had to get out of our comfort zone. And Mike had made sure we could do this by taking us out of the office on tours and missions that had tested us to the limit!'

I'd come from nothing, so why would I think or act like I was better than anyone else?

Martin was right. I wanted to challenge my team by putting them in situations they couldn't control so they could see they had nothing to fear in the end. Then, when they were back at work, they felt empowered knowing they were in a situation they COULD control. And it paid off, the team had a deep understanding of my management approach and I could trust them to run their parts of their business without needing to micromanage any of them.

The recent beauty parade had also put the management team through its paces too and only the strongest had survived. I wanted to find more people to join us at Telecity who had the same attitude – one where they would rise to the challenges they would be facing, and they would never let fear hold them back.

Team talk

I had developed a reputation as being a bit of a maverick CEO because I wasn't your average carbon-copy City guy. I operated very much on instinct, and I had very open conversations about fairness and opportunity for all. I'd come from nothing, so why would I think or act like I was better than anyone else? I had strong values, and I made it very clear to everyone in the company that we should all be fair to each other – not just the executives, everyone.

And that's why people stayed. When shit got real and we'd had to fight for our business to survive, my team was always right behind me because they weren't just there to earn money and pay their bills. They were emotionally involved with the company. Martin explains what that felt like to him:

> 'Everybody got really emotionally involved because Mike led in a way that I think very few managers do, and that is by extreme example. From my perspective, Mike never expected anything from anybody where he didn't expect 120% of that from himself.'

But in every business there is a time for change, and for us that was at the beginning of 2007 when we merged.

Dutch courage

Before we set to work as the new TelecityRedbus, we needed to consolidate some of the Telecity/Redbus roles and put the strongest people in the places where they could have the biggest impact. The day before we sealed the deal with Telecity, I was deciding who would be the best candidate from the existing Telecity/Redbus employees to run each country. I had agreement from the board that I would have control of who did what.

Most of the decisions between the current leaders of Telecity and Redbus were straightforward. A decision only needed to be made where roles overlapped, and it was generally easy to see who was the better manager for the role. But in the UK, neither party had a strong leader, whereas both the Netherlands subsidiaries had strong leaders. I had been really impressed with Alexandra Schless who was running Telecity's Amsterdam business, and I thought she could do a much better job of running the UK operations than both the UK MDs were currently doing. The UK was a much bigger job, so it would be a step up for her, but I knew she'd be perfect. But when I asked Alex if she would like to take the UK job she said, 'No, actually, I don't really want to leave Amsterdam.' So I decided instead to offer her the Amsterdam MD role, which she took.

I didn't want to lose Alex, but I also didn't want to lose the Redbus Dutch manager, Adriaan Oosthoek. He was a great guy, who exuded integrity and thoughtfulness. He was very straightforward in an 'accountancy' kind of way and very good at what he did. He was a believer, and he was prepared to deliver on the strategy.

So I was out having dinner with Adriaan near the Red-Light District in Amsterdam and I was planning to ask him the same question I had previously asked Alex about taking on the UK role. Then, all of a sudden, I get a call from Ian Nolan, Chief Investment Officer from 3i. I went outside into the street to take the call and he told me, 'By the way, you're absolutely going to choose Alex to run Amsterdam.' He didn't know that I'd already agreed that role with her.

What I didn't know then, but found out later, was that the 3i side clearly didn't trust me and thought I was going to get rid of Alex. They'd thought I was going to shaft her somehow, and that story had percolated all the way up to Ian without me knowing. They must have thought I was going to surround myself with my Redbus team and they knew how good Alex was so decided to step in.

So, I'm in the middle of having dinner with Adriaan, trying to convince him to come to the UK and now I am being told what to do by 3i. I'm pissed off. I said to Ian, 'The way I understood that this was going to go was that I was to get complete freedom to make all the decisions about the business.' And Ian said, 'No, this is the way it's going to be and you are going to listen to me.'

I hung the phone up and I called Oliver Grace, one of our original Redbus non-executive directors, and I said, 'Do you want a verbatim of what I've just heard from Ian Nolan?' I told him and he hung up on me. Two minutes later, I had a call. I had Glenn August, Boris, Bo Bendtsen, Oliver, all of them, on the other end of the line. Oliver told me to recount again what Ian just said.

I said, 'At the end of the day, I'm very happy with Alex in Holland. But I thought we had agreed I was in charge and I wasn't going to be told how to run the business or what to do here?' Then, they told me to get back to the UK to meet them at four o'clock tomorrow afternoon in London.

They called 3i and told them they were all getting on planes to London, and that the deal was off!

This was the eve of the deal closing. All the documents were ready for signing. But they meant it. The deal was off if I didn't get control.

When I got to London the next day, I walked into Ian's office in Waterloo East. They were all sitting down around the table, and I was the last one there, they had been talking for ages.

Glenn said something like, 'Hi Mike, c'mon over,' and I sat between Glenn August and Ian Nolan, and I just thought, 'Oh shit. This is going to be embarrassing for someone!' And Glenn said, 'Ian's got something to say to you.'

I don't just hire based on experience, I'm looking for a spark and potential

And to my complete surprise, Ian said, 'I'm really sorry. I should never have told you how to run the business, you have complete control. Do whatever you feel is fit. Apologies, again.'

And then they said, 'Ian, can you just go and wait outside,' and, like a naughty little schoolboy, he did.

They turned to me and said, 'Is that okay now? We're back on?' And I said, '*Look. I'm happy.*' And Boris and Glenn said, 'If we're quick, we can still get to Battersea and get the helicopter out to Wentworth and get a round in before we have to fly out.'

Ultimately, I convinced Adriaan to run the UK operations for me. I said it would be temporary, but he was to do a fantastic job for me there over the next seven years. Adriaan now works with Interxion. Alex currently runs North C, a combined data centre operation in Holland where I am chairman.

The Cable and Wireless guy

Personally, I don't just hire based on experience, I'm looking for a spark and potential. Throughout my career, I have found that it is far better to recruit the right attitude and train skills, rather than the other way around. So, once the dust had settled and I was looking for a head of media, I was keeping my eyes open for someone different. I didn't have a huge amount of time on my hands to interview people, so I asked the recruiters to narrow down a list of candidates. They sent me two people: one worked as the CEO for BBC Worldwide, Julian Turner, and the other was a network engineer and project manager for Cable and Wireless, Rob Coupland.

They were clearly very different candidates offering a vastly different set of skills. The role was more marketing than operational, and I employed the guy from the BBC. He was brilliant for the role, and we got on really well in the interview. (Incidentally, we subsequently learned that Julian was a fantastic singer and guitarist, and a couple of years later he was to get our team thrown out of a restaurant in Estonia after a management meeting for breaking out in drunken song!)

But I didn't want to let the other guy go. Sure, he was lacking experience in the area I was recruiting for, but I saw qualities in him that I knew were invaluable. So I told him I wanted him to come and work for me. Slightly baffled, he asked me what the role was, and I had to tell him I didn't actually know. He then asked what the salary was. On the hop, I told him I'd match what he was currently earning.

Of course, he said no. What I was proposing involved a longer commute for the same pay and no real job title. Not many people would jump at that mysterious offer! I tried to tell him that he would be joining a special team, with a unique ethos… but Rob wasn't having it.

By chance, it was the lead-in to Christmas and I had all the European managers over in the UK and we were having a Christmas dinner around a management meeting. So I convinced Rob to come for some Christmas drinks and dinner with me and the team just to mix and try and get a sense of the ethos of the company.

I met him at Asia de Cuba in St Martin's Lane, London, where I was hosting the evening, introduced him as my friend and then left him to it. I didn't see much of him that evening. But I guess he chatted to a few of my colleagues, built up a sense of who I was and what I did, and sussed out my unusual proposition.

He came back to me the next day and said, 'I'm in. I get it and I'm in.'

So, he started working for me. On the first day, I got him to do some bits and pieces with our networking systems. He got stuck into the task at hand brilliantly. I could see he was working well, enjoying the company and getting to know the ins and outs of the business.

In February 2007, after just two months in the company, I asked him to come to see me. I told him to stop what he was doing with the network and help me on something else. We had decided to IPO the company and I wanted him to run the IPO process. He was gobsmacked! He didn't have any financial experience; didn't know about investing, banks, or anything about the process of IPOs for that matter.

I said, 'Look at it like this: you're a project manager. Deliver me a project called "IPO the Company."' So off he went to start his new project, with a puzzled look and a bit of head-scratching...

But he did it! By October of that year, we had successfully IPO'd the company. We were the last tech stock that would be IPO'd in London for a good few years due to the financial crash. We had Deutsche Bank and Citibank in on the process, and they commented that it was the smoothest and most successful IPO they'd ever participated in, all thanks to him.

The IPO in October marked a significant time in my life for more than one reason. I had met a girl who worked for the London Stock Exchange earlier that year and started an affair with her that would ultimately end in my divorce. Yes, I was a relationship low-life, always looking for that extra ego boost, fundamentally down to a loveless childhood and an abusive father, but that's all for another book. Suffice to say, it took my amazing wife, Shalina, to get me onto a path of feeling truly loved.

So there we were, the last IPO of a tech company for some five years after, thanks to the collapse of Lehman Brothers Holdings in 2008. Our IPO was scheduled to be sandwiched between the aborted IPOs of Smartstream and Sophos, but unlike them, we decided to go for it.

Our successful £436m IPO was on the London Stock Exchange. The shareholders took amazing profits and the company raised net incremental proceeds of £96m, of which £71m would be used to de-gear our balance sheet, and the rest would allow us to move forward with an investment plan that included multiple data centre acquisitions and expansion across Europe[43].

Trust your instincts

Earlier in 2007 we'd said goodbye to Carl Fry and recruited Brian McArthur-Muscroft as the new Group FD. I didn't like him much. Few people at Telecity did. He was probably too much like me! We, like most businesses, had been hit by the 2007-08 global financial crisis, the biggest economic downturn since the great depression of the 1930s. With global businesses falling into debt and bankruptcy, we were at risk of losing customers, and with the added risk of not getting an IPO away at all in the middle of the financial crisis, I needed a slightly less risk-averse CFO than normal. Brian was definitely less cautious than most CFOs I have met, but he got us through the IPO. The company was still coming out of its challenges and consolidation, and reps and warranties (which are the assertions that a seller makes during a transaction and are relied upon to be true and correct) needed a liberal approach. Most of the hard work was done by his under-studies David Crowther and Brad Petzer. The latter is now my CFO at Pulsant, where, incidentally, Rob Coupland is CEO.

I'd also hired John Hughes as our new Chairman from a shortlist provided by Deutsche Bank. He was another big character. He wanted to be a FTSE 100 chairman. He wore Berluti handmade shoes and had an office in Mount Street with F1 memorabilia, including an F1 racing car wheel as a coffee table. He claimed to be best friends with Bernie Ecclestone. I always had a gut feeling niggling away about John though, and my instinct would turn out to be right. But more about that later...

PART 3

COMING OF AGE:

Work hard, play hard

CHAPTER 7

RIDING THE ROLLERCOASTER

The only way is up

The years that followed were filled by some of the highest and lowest points of my life. I was going through my divorce and struggling with everything between 2008 and 2010. I started drinking a lot. My brain-to-mouth filter had somehow disappeared, and I was less and less able to be diplomatic. I suppose it's unrealistic to think that the pressure over the previous six years would not hit me at some point, but looking back, that version of Mike Tobin was not who I was, or who I wanted to be.

Outside of my personal world, the rest of the world's reliance on technology was growing minute by minute. Music, video, and

"

Google was no longer just a search engine; it was a verb

smartphones were driving web activity to at least double every year. The first iPhone in 2007 had sparked a wave of mobile-friendly websites, which continued over the coming years. Print was dying and people were seeking instant information online. Google was no longer just a search engine; it was a verb.

We were in a new age, one in which the internet had planted itself firmly at the centre of our universe. It was the primary source of 'at a glance' information and communications, and we saw social media, eCommerce and email explode.

In 2009 Twitter reached a valuation of $1 billion and Facebook added over 200 million new users.

In 2010 Pinterest and Instagram used photos to take the place of words, and apps came into their own, with 35% of adults using them by the end of that year. Ninety percent of Britain's internet traffic was passing through our data centres, and the need for more and more data centres in which to share, exchange and distribute all this electronic information globally was profound. The formula behind our growth strategy was simple – the more people used the internet, the more capacity we needed to facilitate their use. They weren't going to use less tomorrow than they did yesterday, so we were heading in one direction – up!

Sales were up, profits were up and share price was up… it felt like the world was our oyster.

In 2011 we paid out our maiden dividend to shareholders. When we had floated in 2007, shares cost 220p, now they were changing hands at 648.5p. Our revenues had grown 22% in the last year to almost £240 million, and profits were up by 29% to £59 million.

We were the shining star in the European data centre indus-try, and it felt good. I'd had plenty of rotten eggs thrown at me along the way up, so I was happy to enjoy the praise we were getting now! During one set of results, we celebrated until 7am and then went straight into a meeting with the press, but I couldn't do the presentation in front of shareholders (Brunswick offices) so we gave it to a colleague and exited out the celebrity back door as I felt so sick.

Hitting the road

Since 2010, the data centre industry had been exploding. We were the darlings of the new economy, with a booming industry that showed a share price only going in one direction. To keep our profile up and show potential investors and the rest of the world what we had to offer, we used to head off to the US on industry roadshows.

These were conferences that many of the banks would hold in various cities. You'd get to the conference hotel and be sent up to a room, which had its bed and all the furniture removed, and one round table put in the middle of the room. You'd sit there with eight or ten chairs around you and investors would come and 'speed date' with you.

So, on investor roadshow days, you'd have 40-minute meetings with different investors or prospective investors from seven in the morning to seven at night. They would come in and ask exactly the same questions every single time. It was a bit like being on a press junket for a new movie release.

It was a long day. You weren't allowed out. Instead, we would be brought these little sort-of happy meals with a sandwich or something around lunchtime so that people ate while they were having their meetings. And then, every now and again I would say, 'Could you just hang on a minute?' And I'd nip to the loo in the hotel bathroom. That was literally the only break you could get.

And then, if you were really popular, you ended up having pre-pre-dinner drinks, pre-dinner drinks, dinner, and then post-dinner drinks with other investors as well. In which case, you are on the go for almost 24 hours (which, from six or seven o'clock in the evening were fairly alcohol-fuelled).

It was an amazing time, but most days were banal because you were just answering exactly what investors wanted to hear, and you'd forget who you had and hadn't told what, because you were just saying the same story 20 times a day. But, outside of the meetings, the events you could go to were really good.

There was one particularly funny day when I was on tour. The head of Jefferies TMT (Telco, Media and Technology) team had rented a convertible in Miami and was in the driving seat. Like everyone back then, they had been given a very flashy car to keep themselves mobile during the conference period. There were four seats in the car, and a senior manager sat in the front. I was in the backseat with Byrne Murphy, founder and owner of Digiplex in Norway. We were standing up on the seat, waving our belts around, as we're driving down South Beach, with the guys in the front just petrified they're going to get arrested and thrown in jail!

Next thing they know, I've lost my belt! It flew out of my hand onto the sidewalk.

My great friend and top industry analyst, Milan Radia, was a broker at the time with Jefferies and would often also be on the roadshow tour. He called this time in Miami the 'gold rush' era:

'Jefferies started off having TMT roadshows in New York, but in the last few years, we were in Miami in an effort to kind of try and glamorise the event a little bit and get people to stay a couple of days. We had good meals, good events and it was a good time.

Byrne told us the belt-waving story and said to me that the following year he wasn't going to be seen in any motorised vehicles with Mike again that was for sure.'

Well, what can I say, I always made an effort to make a lasting impression!

Another TMT conference I remember was the JPMorgan conference in Boston. I always used to stay at the Liberty Hotel in the centre of town. It was a converted jailhouse and the room key used to look like a cell door key. A great place to stay. The conference was always at the Westin Waterfront Hotel, about a 20-minute walk away. My contact at JPM at the time, and a really good friend, was Rupert Sadler.

On the eve of the conference, Rupert's boss, a certain Fred Turpin, who I never really liked much, hosted a private dinner for a dozen or so CEOs from the industry. I sat next to Steve Smith, who was the long-standing CEO of Equinix. Despite being head-to-head competitors we always got on well, although we only used to catch up at conferences, or when Steve was in Europe on a trip.

During the dinner, the alcohol was flowing fast and it was getting pretty animated. I was joking in my usual larger than life fashion, which is a little disconcerting and a bit alien to many of the big US company CEOs, and I turned to Steve and made a comment about his cufflinks...

The week before an institutional investor in both Equinix and Telecity had been talking to me about Steve and said that he always looked at the flamboyancy of Steve's cufflinks. If they were bling, he knew the business was going well. It was a bit like me and the buttons on my shirt! Anyway, Steve took offence at this and shouted at me 'who do you think you're talking to? I am the CEO of a multi-billion-dollar company and I deserve more respect!' I just burst out laughing, stood up and said, 'I am done with this crap,' and walked out on the group. Everyone was silent and I am sure I became the topic of conversation for the table as soon as the door was closed!

On the walk back to the hotel I texted Steve, 'I can't believe you've turned into such a dick!' and waited for the aggressive response. Eventually, it came, 'I am so sorry you feel that way.' Uggh! I felt so bad. He had made me feel bad by being the gentleman!

A couple of years later (yes, it took that long for us to speak again), after Telecity had sold to Equinix, I caught up with Steve for lunch in London in an attempt to bury the hatchet. He explained that he couldn't even remember the episode as he was drunk that night and wondered why I hadn't been in touch!

He is such a great guy and as fundamental to this industry's journey as anyone else in this book.

In January 2018, at the height of the 'ME TOO' movement, Steve stepped down as CEO of Equinix after 'exercising poor judgement with respect to an employee matter'. Well, we all make mistakes... you can find Steve now firmly back in action in the industry as Managing Director of GI Partners, the private equity company originally behind the launch of Equinix's biggest global competitor, Digital Realty!

Viva Las Vegas

Another big industry player at the time was DuPont Fabros, a conglomerate who had originally operated in chemicals. They moved into data centres when Lammot DuPont, the last-standing heir of the DuPont family and an analyst for JP Morgan, partnered with Hossein Fateh of the 'Fabros' Fateh brothers – who were real estate developers in Washington. Together, they started buying up data centres that belonged to defunct Internet Service Providers. It was a shrewd move and they fitted in perfectly during the heady Vegas days of casinos and private jets.

Hossein once invited me to fly back from Vegas on his private jet. We were there for an industry conference. Unfortunately, I couldn't make it as he was going home to Washington and I was going to NYC. On the way out of a conference, as we were chatting and walking through a casino, he abruptly stopped and decided to make one quick bet on a roulette table.

He basically threw his money on a number, and it came up! He then just took the chips and literally just walked away from the table and said, 'I'll be back in a minute.' Everyone thought he was going to the loo, but he went straight to a Rolex shop and bought a Rolex watch! Because state tax is quite important on an expensive item like that he came back and said, 'I've just bought a Rolex tax-free.' And he wore it back on his plane. And that was the sort of guy he was! And that was how the industry was at that time...

That relationship with DuPont Fabros was important to me too. Hossein was a true pioneer in the industry who'd made pivotal improvements to the way that pricing was managed, which became very attractive to customers and helped data centres to grow worldwide. His unique achievements were demonstrated in 2017 when DuPont Fabros Technology signed the biggest-ever data centre merger deal (at the time) with rival Digital Realty for a whopping $7.8bn. According to Vinay, they had just 34 clients and around 115 employees – giving them the most valuable revenue per employee ratio ever known in the industry. Vinay remembers having an interview with Hossein the following week and being shown the same prized Rolex!

Buy. Buy. Buy.

Back to 2011, and the industry was moving fast. Telecity were well-established in London, and expanding in Manchester, enabling the city to emerge as a new technology hub. Between 2008-2012 we'd continued to buy up existing data centres across the UK and Europe, including Dublin, Bulgaria, Istanbul and Poland – the latter of which was unusually located at

**THE TOP OF THE MARRIOTT HOTEL IN WARSAW - HOME OF PLIX,
THE POLISH INTERNET EXCHANGE**

the top of the Marriott Hotel in Warsaw. I remember having to walk through the hotel staff washing machine area to get access to the top floor of the hotel, where the data centre was, with Sylwester Biernacki who became my Polish country manager. We had acquired a business called 3DC in Bulgaria, and nearly bought another which was linked to the local mafia. Fortunately, my Bulgarian friend and country manager Zdravko Nikolov did a bit of detective work on the company we were looking at and found out its true roots just in time for us to abort the transaction.

The data centre we bought in Ireland, Data Electronics, was owned by a consortium of investors, including the charming Gary Hersham, who owns Beauchamp Estates in London, and Edwardo Azar. He was a great guy and invited Shalina and me to his daughter's wedding in Argentina shortly after we bought the business.

Eduardo is a Jewish Argentinian, who was living in London at the time, and now lives in the Caribbean. The deal was progressing, and as we approached the last details to be nailed down, I decided to call him to make sure we were still on track to close the deal on the Friday of that week. He confirmed this and said he would be travelling on Friday. He was going up to Scotland to play golf at St Andrew's.

Friday came and, as always happens with these situations, there were still finer points to resolve at the final hour. It was winter and the days were getting shorter. At lunchtime, I called Eduardo and asked him where he was exactly so I could work out the exact time of sunset for him so that I knew what the cut off time would be for him to be able to sign the contracts – as a practising Jew, I knew he would not do a deal on the Sabbath, which starts at sunset on Fridays.

He told me later that he was so touched by the fact that I was sensitive to this detail, he signed the document in advance and sent it to the lawyers before sunset, even though we had not finalised the agreements. He said that he knew he could trust someone who was that thoughtful towards him. Another example of how thoughtfulness pays in business.

As part of the transaction I also 'inherited' Maurice Mortell, the MD, who is still the MD of Ireland for Equinix today, and who, along with his wife, Jacinta, became good friends to Shalina and me.

Run, run as fast as you can

Generally, our expansion was well received, but with the growth and the fact that supply and demand was finally in our favour, we also had started to increase our prices at a significant rate; 60% across the board one year, 25% the next. Some of our clients were far from happy! Some clients left, but most appreciated the fact that they had enjoyed cheap pricing over the years when times were tough for us. We were simply introducing pricing that allowed the business to make enough of a return to justify reinvesting in more capacity for the future. We were followed pricewise by the rest of the market. Not all our clients were so amicable about it. A certain Lawrence Jones had founded UKFast in 1999 with his wife, Gail, to provide web development services. They were based in Manchester, with hosting now provided by Telecity. When we doubled our prices for UKFast, Lawrence went ballistic. He behaved like an utter prick.

He protested, wrote letters of disgust and insisted he would leave if we didn't backtrack on the pricing. We didn't, so he moved to another data centre called Internet Facilitators Limited (IFL).

We had been looking at the Manchester market for some time and had our eye on IFL, and when it came up for sale just a couple of months later we snapped it up. No sooner had we taken ownership and started hanging the new logos on the building, we doubled the prices for the IFL customer-base. Including our friends at UKFast. Lawrence was really not happy this time. In fact, he was absolutely livid. He was sending me hate mail by now. His emails were aggressive and threatening, implying that we were taking the food from his children's plate and that if we didn't reverse the increases he would 'make it very bad for me!'

Again we took no notice, and poor old UKFast couldn't move fast enough! Each time they moved it was a significant upheaval and disruption to their business, but they operated on tiny margins so they had no choice. Next, they relocated to another data centre called UKGrid, the last independent data centre operator left in Manchester... which we also later bought! Corresponding price rises followed and so did the threatening letters from Mr Jones. I alerted the police to the threatening nature of the correspondence but nothing further evolved.

UKFast eventually raised investment and moved into their own purpose-built data centre and we heard no more of them.

UKFast has since been successful by all accounts, but as I write this, Lawrence is at the centre of a very public investigation into wide-ranging misconduct in the workplace[44]. As the press reports it, the 'darker side of the British tech baron' has become the subject of the investigation.

The race to the top

With all this buying, it might look like our growth and dominance of the data centre industry was easy, but in reality, it was not easy to thrive in an era of austerity.

Virtually all the markets had shut after the Lehman's disaster, we were in a recession and businesses were going bankrupt around us. The economy was bulging with players who'd flourished in the heady boom of the mid-late 90s but were now wasting away thanks to the drought that had followed. We were in a very unique place. Our growth was being fuelled by the unstoppable rise of the internet, but on the other hand, we were not a technology company ourselves. We were, in fact, a glorified real estate operator, a server 'hotel', which housed servers for businesses that needed to store and share data. But real estate was not multiplying at the pace people needed it to. We were providing a physical space, with a lot of power allocated to it, which is a scarce resource, being filled with businesses that grow on the back of online activity. It was a seemingly endless growth curve…

The industry was also developing at speed across the world. Regions like South Africa that had previously relied on America and Europe to provide their connectivity were now emerging as competitors by building their own data centres. The sooner we consolidated the sector, the better.

But there were two big players who were dominating the global market; Digital Realty and Equinix. There was one other European operator, about the same size as us; Interxion.

Milan Radia explains how he feels we were able to stand out and give our competitors like Interxion a run for their money:

'Data centre companies had their work cut out after the IPO, and Interxion had to spend some time really getting to know investors, with Telecity having the advantage of having been listed for some time.

Even prior to the merger with Redbus, Mike and both businesses had been followed and understood by investors for a number of years. So Interxion coming out of the blocks with their IPO in 2011 had to start meeting with investors later that year. It was a European company listed in the US, so inevitably at London meetings, the investors were very much saying, "We know Telecity, we know Mike, who are you guys?"

So, as you can imagine in investor meetings, it can get a little bit frustrating when you're having to kind of overcome the cult following that Mike had created in Europe, certainly the UK.

I don't think we should overlook the importance of how high Mike had set the bar for other people to break into the industry in terms of getting investors on board. There were a lot of positive feelings being generated towards him, and it helped that, along the way, performance was doing pretty well.

Some CEOs are run-of-the-mill, and they'll interact with investors on a very normal, relatively functional basis. I think what Mike did well over the period was manage to find a kind of balance where he created a little bit of magic *by showing his real self. He has a colourful persona who could just basically turn up to a meeting with his business card… he'd then proceed to spend 10 minutes showing us how an image on it pops up from a QR code.*

So what you have here is this charismatic guy with a load of shirt buttons undone and jewellery on display, and it was very different from your normal CEO meeting. Entertaining and actually delivering, because these investors were not going to take entertainment value without a delivery. You still need to show the right financial performance. We found that for a prolonged period of time the combination of the two created a very good performance for Telecity because it gained the investors' support.

Mike talks about the Roadshow conference schedules being 20 consecutive slots a day, and that's day one and the same again day two. But that actually wasn't the same for everyone. There were companies that were extremely popular and Telecity was one of them. The investors all wanted to be around them – they had this persona that was courageous in terms of Mike's external-facing performance, and around the company, which was actually beneficial to the way investors looked at them.'

Game over

We'd recently acquired two companies simultaneously in Helsinki; Tenue OY and Academica OY. We bought both because in secondary markets you need to be the 'game over' number one.

Whereas in the primary markets you can be a strong number two or three and still make money.

When I flew to Helsinki to meet both the target management teams, I happened to notice a couple of employees from Equinix were on the very same aeroplane as me. I knew they were meeting with the same companies who had already indicated their desire to sell. I couldn't afford to be one of two players in the Finnish market, especially if the other one was Equinix, so we bought both at the same time and shut out Equinix and all others for good.

What we really wanted, however, was Interxion. The pan-European player was led by the infamous CEO, David Ruberg, and we wanted to buy them. Ruberg was having none of it.

Interxion was founded in 1998 in the Netherlands, so we'd been competing with them for years and I'd tried to do a deal with Ruberg multiple times. The BOB team and I met with him on many occasions before our IPO – at the last one he had started the meeting with, 'Why should I sell to you?' and Bo Bendtsen got up and said, 'We are wasting our time here,' and walked out. The meeting had lasted four minutes.

I told him he was a loon to his face. Being American I am not sure he understood. I didn't like him at all, but he mellowed with age and had a prolonged illness by all accounts. I have heard from people who have worked with him that once you get to know him, he is quite a good guy, but at the time he came across as obnoxious and very difficult to engage with. We always seemed to be the darlings of the industry and they were looked at as less innovative, less exciting, and perhaps a

169

little boring. Interxion later became fundamental in the rest of the Telecity story, however... as you will read later.

I bumped into Ruberg again in 2019 in the concourse at Schiphol airport. I had just arrived and he was heading out. I stopped him and said hello. He said, 'Hi Mike. Long time no see, I am almost too afraid to ask you what you are doing in Amsterdam...' I laughed and told him he was right to be worried... but what I didn't tell him was that I was working with DWS Private Equity and we were in the process of buying TDCG and NLDC, two Dutch data centre companies that are now merged as North C Data Centres, one of his biggest competitors on the Dutch market!

Leading by example

My view with Telecity was that we should continue to grow, but that ultimately, we should do whatever was best for the business with regards to shareholder value, whether that meant we bought, or we sold. I was always objective in this way and not so blinded by ego that I would obscure that principle. The single reason I was put in the role as CEO, was to be the guardian and creator of shareholder value.

I could see that Equinix and Digital Realty had more capital than we did, and I would have been prepared to hand over the keys if a deal was right for our shareholders – but *only* if it was right.

When it came to Interxion, we were of similar size, and a merger of sorts seemed the most logical goal to pursue.

I'm told now that the difference between my leadership and that of others was my willingness to spend time talking to

"

The difference between my leadership and that of others was my willingness to spend time talking to people

"

people, building up relationships, and even having fun. Which, in an acquisition process, is what helped me to work out who we would have a good connection with, and who we wouldn't. It also helped us win competitive situations, even when other bidders were offering a higher price.

I learned what others thought of me as a leader from my dear friend Joseph Valenti recently, as he recalled what it was like working in the industry as a banker for Lehman Brothers and Barclays in New York:

'In any business, you have the ability to connect with people, the ability to get people to trust you, to open up to you, to see you as something more than just being one of multiple facets to their life and what's important to them.

To be honest, there's a lot of people who sit in Ivory Towers thinking, "Okay, I've now kind of arrived and I'm going to be extraordinarily selective around what I do, how I spend my time, who I spend it with..." But your willingness to spend time with them, engage with them, tell them what you know, took away the dread of having conversations about what's going on in the business, the industry, where there are opportunities, what you are seeing, what we are doing, and how we could do it better.'

Managing all these data centres was not easy, at our peak, we had 48 and each one of them had to be managed well in order to achieve its potential. It was down to each of our country managers to make that happen, but I'd fly in to help them deal with all sorts of issues that cropped up over the years, from legionella scares to fires, and floods to electricity outages.

Black Monday

Back in 2005, in the Redbus days, we'd had to cope with what became known in France as 'Black Monday' when the entire internet went down… and it was our fault!!

Our French data centre had basically just crashed. It was a Sunday in February, and the centre was managed by Stephane Duproz, our very able and loyal country manager. He remembers the fear as he rushed to the site to find out what had happened:

'I looked out of the window, and in front of the entrance, there were about 60 people, journalists, newspapers and customers. The customers were angry and we were in big trouble. They were extremely, extremely unhappy. You know how French people can be when they are extremely unhappy!

So, I decided, the only way I can get out of here, really, is to go and talk to them, and okay, they are angry, but surely they can be reasoned with.

It was very dangerous. I went down to talk to them, and they wouldn't listen, and then I also started to be extremely unhappy, and it was very close to being physical. One of the people who commented afterwards said, "You were very close to being physically attacked." But I just stayed there, because there was no other way. We had to explain what we were doing, and to communicate with them.

We made several attempts to get back to the normal situation, but our attempts failed. It just happened that at the time we were connecting the website hosters for most of the country, so a lot of websites were down.

The hosting companies didn't have a disaster recovery plan, they were just simply, perhaps irresponsibly, assuming that a data centre never goes down, so they were really left in a very bad situation. But in order not to appear as not having done their job in having disaster recovery capacities themselves, they told their customers that it was our fault... So, our own customers told journalists to communicate with us on a real-time basis on what was going on, meaning all eyes were on us rather than them. It was very tough for me to take. Very, very tough.

This, of course, was all because Redbus, and all the other data centre operators at the time, were so strapped for cash, we just didn't have enough money to maintain the infrastructure and put in the right resilience. We were just praying nothing would happen!

As it transpired in the independent investigator's reports afterwards, our maintenance was good, it was clean, and all the coolers worked. It was just a matter of power factor. At the time no one knew about power factor, but now thanks to the recent BA data centre outage, we all do!'

Que sera sera

In fact, our French data centre went through quite a few traumas. For a year and a half, we had an illegal travellers camp that had set themselves up right opposite our site. It wasn't a great look when we had the heads of businesses visiting us to decide if they wanted to use us, so we had to fight to get them removed.

One day they were finally told to move on, but they were one step ahead and realised that if they burned all their belongings the council would have to find them somewhere else to live instead – because in France when you are a victim of fire you have to be provided with shelter.

So they started an almighty fire.

Stephane remembers the event:

> 'The fire quickly spread and set light to an adjacent warehouse, in which there were plastic shoes, you know, trainers, in cardboard boxes, and it started to be extremely, extremely, extremely big, and it was something like 25 metres away from where we were.
>
> So, obviously, police came, the fire brigade came, and they started to enter the site, discussing what to do next. Then a big policeman came to me and said, "Okay, you are evacuating the site." I said, "No." He said, "What do you mean, no?" I said, "I mean, no. I can't do that. I have customers in there."
>
> So we evacuated every customer, evacuated the salespeople, evacuated everybody. Then the policeman came back minutes later, saying, "Okay, now you will evacuate yourself too." I said, "No." He said, "Okay, I will come back."
>
> The guy comes back 10 minutes later, with the head of the fire brigade. The head of the fire brigade comes over

to me and says, "Now sorry, you just really have to evacuate." I continue saying, "Absolutely not." At that point, the policeman started to physically intimidate me. He was close to two metres tall, and he really wanted me out, and I said, "No." It became physical. Then, in the end, he said, "Okay, let me come back."

He came back 10 minutes later and he said, "All right, only absolutely indispensable people can stay." So, there was about six or seven of us guys left, and I told the guys, "All right, who is dispensable here? You have to understand that we did not know when we would be able to leave. Maybe we would have to stay for two days. So, if you want to leave, leave. If you are indispensable, stay."

Nobody moved. Everybody said, "I'm indispensable."

We ended up leaving the site something like 18 hours afterwards, and it became very, very, very tense. The flames were virtually touching the building.

This was an amazing display of loyalty in what was a truly life-threatening situation. Thankfully, however, everyone was fine and the site was safe. Just like John Polak, who withstood the floods in Prague, our team was fiercely loyal. Not because we asked or expected them to be, but because they were genuinely and wholeheartedly connected to the business. And for that, I cannot thank them enough.'

Another time, I think at our French site too, we had to investigate a potential legionella issue with the water-cooling system. Water-cooling systems are necessary in data centres because the computer equipment generates a lot of heat, and that needs to

FIRE ON OUR DOORSTEP IN PARIS. THE DATA CENTRE IS ABOUT TO BE ENGULFED IN DEADLY SMOKE!

be cooled to keep the servers at the correct operating temperature. The water-cooling towers can be dangerous because water evaporates from the tower, and the legionnaires' bug can live in the warm water. If it is not killed it can dissipate with the vapour and cause serious harm.

We followed protocol and got the authorities involved, but it reached crisis point with us constantly having to turn the coolers on and off. Our staff were afraid too and we were trying to find the right, sustainable solution.

Eventually, we got there and it was a happy ending, but I'll never forget that after all the technical tests had been done and we got the all-clear, one of the final tests for the water quality was to put a couple of live crevettes (prawns) in the water, and if they were still alive after 48 hours we were good to go! I remember thinking it's just always about food for the French!

I would see Stephane once a month for a review of the French business. The usual way was I would get the first Eurostar to Paris at around 6:30am, arriving around 11am into Gare du Nord. Upon arrival, I would walk out of the station and across the road to a small brasserie called Terminus Nord. It's been there since 1925 and is a classic Parisian brasserie and traditionally open 24 hours a day.

We would begin with a glass of champagne and the starter would always be a large plateau de fruits de mer pour deux. We would discuss the business performance, the outlook for the future and any opportunities or challenges the business faced whilst cracking open langoustines and downing oysters. More often than not, I would order Sole Meuniere for a main course and Stephane would highlight his love for Burgundy white wines with a bottle of Chablis, a Montrachet or a Volnay or two. We would then discuss more emotional issues – HR challenges, personal and family issues. I found this to be very important. The role of an MD is a lonely one and people need to have an outlet to be at peak performance in their work environments.

Then, at 4pm, I would get up, bid my farewell and walk across the road and get on the 16h30 return Eurostar to London.

I visited every country in the group every month. Many were day trips but most were overnight and the informal dinners with the country managers were always a key element of how we got the best performance out of the team.

CHAPTER 8

PARALLEL LINES

The IXEurope Story

We certainly weren't the only ones to endure the rollercoaster ride that was the ups and downs of the data centre industry from the late nineties onwards. My good friend and colleague, Guy Willner also experienced the crazy rise and fall of the industry, as the founder of IXEurope.

Back in the late nineties, while I was still working overseas and data centres were not even a glint in my eye, Guy worked for General Cable in the UK. He spotted a gap in the UK market for a new data centre when there was a 30-minute outage at Telehouse – the UK's one and only data centre at the time, housing 95% of the UK's internet – the country came to a stand-still.

Over the course of a decade, Guy and I followed very similar career paths, although he eventually sold his company to a much bigger beast, whereas my career took a different turn (as I will later explain).

We are both known in the industry for our storytelling, and I've explained how Redbus Interhouse and Telecity arrived and emerged to date, but there was a third player on the scene, and that was Guy's baby, IXEurope. I have invited him to tell his personal story here to share what the industry looked like from his point of view, and how our lives were running in parallel lines whilst we were all battling it out to be the top of the industry.

'Even in 1997, when no one really cared too much about the internet, it had already become vital for many businesses. So when the whole of the UK internet went down for so long, I thought, well, in 10 years' time, Goldman Sachs is going to be doing some foreign exchange transaction on the internet and if it all falls off for half an hour, they're going to lose $50 million. Maybe we need a second data centre in the UK.

I was working for General Cable at the time, but I thought, "Why don't I try and do this?" First, though, I'd better try and understand the financials of data centres. So, I telephoned Companies House in Cardiff. Because in those days you couldn't get any documents on the web. I paid two quid per document and they faxed me through 10 years' of accounts for Telehouse, London. Telehouse had been founded and built by some visionary Japanese companies back in 1988, partly as a disaster recovery site in the unloved Docklands for a number of major Japanese banks. One of the "IT suites" was reputedly decked out complete with oak panelling. When telecoms deregulation started new operators began asking if they could put their freshly acquired telephone exchanges, sometimes literally waiting in the docks in shipping containers, into the Telehouse as it seemed a good place to house them. (And that is how it all started in the UK.) Poring over the numbers, I saw that by 1996 they were doing about 30% net profit year-on-year, they were just chucking out cash.

I thought it was a brilliant thing! So, we started a brand-new data centre called IXEurope.

There was money around to invest in telecom tech in that space. From November '98 to July '99, we were looking for money. I brought a French friend of mine on board, Christophe de Buchet as co-founder. He had been around the block in business and brought along a guy who'd been an angel investor in his business in the early days. He promptly gave us £30,000 seed money to start the business.

I contacted my old chairman, Sir Anthony Cleaver, who was a chairman of the General Cable Company, which was owned by Vivendi. I said, "Can I come and see you and show you what my start-up is like?"

I went there with a couple of mates. We turned up with our laptops and we showed him some groovy PowerPoints about data centres and stuff like that. He said, "Yeah, I think that could be interesting." I think he thought, "This is fun, a bunch of crazy young idiots doing something completely mad!"

Then he said, "I expect you want some seed funding, do you?" Nervously, we said, "Yeah, yeah... well, yeah!" He

Ask for 20 million and they'll take you seriously!

said, "How much?" We had no idea, so remembering what our angel investor had offered, we just said, "Thirty thousand pounds." He literally got out his cheque book, Sir Anthony and Lady Cleaver, there and then, gave me £30,000, and said, "I expect you'll send me the paperwork shortly." It was a very, very fantastic moment.

By doing that he'd established a level of trust, which is way up there. He was expecting us to be super clean, uber responsible.'

Room for a little one?

'Then, armed with 60 grand, we went to a couple of conferences. I did one in Frankfurt and some market research. From a one-man serviced office, Christophe and I searched for real money.

We thought we needed two million pounds for our first data centre. I started running around all these investment firms and they said, well, two million pounds, I don't know... Then somebody said, "Hey, ask for 20 million and they'll take you seriously!" Initially, I thought that I wanted to build a single data centre in west London – the opposite end from Telehouse in Docklands. Now I thought, well, I need to plan for 20 then. I'll do six data centres in London, Manchester, Edinburgh, whatever... We started looking at it and we thought, well, Christophe speaks French and German. I speak French and a bit of German... let's do Europe – sounds much more sensible! And yes, the funds suddenly wanted to meet up.

Our first big break was meeting Cisco Capital in Santa Clara, who introduced us to an unlikely pair – Al Avery, a serious older tech guy, and Jay Adelson, a young livewire, both working out of a ratty old spare office that Cisco had lent them, as they worked on plans to create Equinix. I was amazed that Jay was already having T-shirts printed with the Equinix logo, even though they had no company yet. Cisco proposed we knock the two start-up teams together, but Christophe and I were worried that we'd end up the poor relations, thousands of miles away back in Europe. So back on the plane.

Shortly after we found an investment fund that agreed to put in £10 million in July 99. First round. All thanks to one PowerPoint! But that was the era we were in – money was being thrown into the new tech economy. That was with European Acquisition Capital. They were not a tech fund, but they were savvy. They had chains of hotels, chains of coffee bars, chains of dentists and they thought fine, let's have chains of data centres. It sounds about the same. It's a cookie-cutter approach, and why not? I think that may well have been the highest-funded first-round tech investment in Europe, it certainly was in the UK. Surprisingly, having a non-tech funder was probably a big reason why we survived the looming dot-com crisis.

Now safely funded we started with a tiny little data centre on the ground floor of an office block on Finsbury Square called London One. It had coolers on the roof, and we had a nightmare whenever we needed to do anything to them as the main roads had to be shut to get the cranes in! (The neighbours weren't so happy as on top of that the street closed practically once a month for a week to bring more optical networks into the building. In their wonderfully non-Cartesian way, the UK authorities never cottoned-on to building just one great underground duct network where all telecoms companies could pull in their fibre cables. No, each telecoms operator had their own ducts constructed, leading to endless road digging, re-opening, re-digging. Great for the construction industry and the legacy is a lot of spaghetti under our feet, mostly in not great condition.)

It worked well so next stop was Heathrow, west of London, where we signed a 25-year lease on a massive (at the time) warehouse just near the runway. I remember looking at this big empty expanse, almost the size of a football pitch with room to stack three London buses vertically, thinking – how are we going to fill this with customers? It was a sobering moment.

Suddenly though we became super-hot. Everyone was talking about the Internet Revolution by now.

A warehouse lease in Paris came next, because we had committed with the investors to do a European rollout. I remember a group of Travellers came, broke through the locked gate and parked around 20 caravans in our car park two days before a big onsite client meeting. There was nothing we could do legally for a 30-day period but Christophe diplomatically persuaded them to move their caravans out of sight for the duration of the meeting!

We were in the euphoria of the dot-com bubble. We did a second round of finance in 2000 and got 42 million pounds – straight in, turning away investors. We had hardly any revenues.

June 2000. 52 million invested. We were really pumped-up – this round was undertaken with a view to float in the company shortly afterwards in October 2000. The whole thing was, "Guys, get moving. You've got to show some momentum for the float." So, we went off and we hired people. We signed a lease in Barcelona. We signed a lease in Amsterdam. We signed a lease in Milan. JP Morgan became an investor. Then, instead of putting money in, they threw in a couple of data centres which they owned and could put into the pot. Suddenly, we had Frankfurt and Zurich in the mix too. We went from two countries to four almost overnight and rolled out teams in all those locations.'

Keeping the faith

'JP Morgan and Lehman Brothers were going to be the float partners. It was another world now. I remember going to a meeting in these really swanky offices and big boss Carlos Condé was there, and they had all their team there, with their notebooks and everything. He said, "Yeah, I'm just the conductor of the orchestra. You won't see me very much. But you're in safe hands. We're going to be doing this, this and this..." They were doing a stack of floats a week at that time.

Of course, we didn't know what we were doing here. I remember a very important moment during our meeting. I was thinking, "Float, float". But I was also thinking, "Shit, it's my daughter's birthday tomorrow, I still haven't got her a present. I know she wants rollerblades." In my head, I was 50/50 on whether to stay in the meeting or go to the shops! What was most important?

After a while, I said, 'I've got to get these rollerblades to my daughter, it's her birthday tomorrow.' And they said, 'No problem, Mr Willner, we can do that. They brought some trainee along and told her, 'You're going to Decathlon to get this 35-pound pair of skates and bring them back here for Mr Willner.'

You got the full taste of what life can be like when things were going well. Flattery was everything and it was a whirlwind. We were superstars. Our investors, our bankers, they treated us very well. We were planning to use a private jet to do roadshows in the US and money was no object.

I was naive at the time, I had faith in everyone, and I thought everyone around me knew more about the importance of everything than I did.

Sadly, not long after, my wife's grandmother died, and the funeral was going to be on the day that we started the float process, day one of all the big investor meetings. The bank was basically looking me in the eye and saying, "You can't go, you have to be here. You can't go to the funeral." I accepted it, but I now know I should have said, "Sod it. I'm going to the funeral. I'm sorry it's a day late. What's that going to change?" That was a very hard day and I remembered it for a long time.

Of course, during all this time, you would sometimes start to worry about all the investment, but the private equity investor is telling you, "Hey guys, don't worry. If the float doesn't work, I'm there. I've got the money. Don't worry, we're taking a bit more risk, we're spending a bit more money, but don't worry because if it flops, we'll prop it up."

And then we were hit. The dot-com bubble burst and it hit everyone. Investors suddenly didn't want to touch us. It wasn't a big BANG, more of a slow fizzling out of all our hopes and dreams for the project. We did the whole road-show in the US and yes, we flew round 10 cities in a private jet, but about halfway through it became obvious the cracks were showing across the whole market. "Sorry guys, the book's not covered," said the bankers, which meant we didn't have enough interest from investors and all those preparations (and legal fees, audit fees, bankers' fees) were up in smoke. Luckily, our original investor was not in the tech industry, they had cafes and hotels and dentists. The dot-com crash was very specific to the internet sector, everyone else seemed to carry on as usual with a "told you so" look on their faces. So our trusty investor was not hit nearly as bad as the many tech funds, and they were able to put their hand in their pocket and keep propping us up.

It did take us six months to convince them to put in a rescue round, however, and we had no visibility. It was like we had a disease that might be catching. We couldn't see more than about three months out.

We then did a second rescue round in 2002 when the fog had started to lift and there was a glimmer of hope; we got about seven and a half million pounds both times.

During this lean time, we tried to exit leases, but this is one of the killer problems for the data centre industry, we had only just signed 10 and sometimes 20-year leases on buildings. So, we had to go back to the building's owner and say, "Well, actually we've run out of money. We are not going to fill this out as a data centre. Can we hand it back?" And they'd say, "You signed a 20-year lease. I'll tell you what, I'll give you two years off. You can pay me 18 years upfront now and you can go away!"

Over the following five years, we restructured, managed to wiggle out of leases, and our share price had crashed to a very modest valuation which was really, really painful, but we got through it somehow.'

Sell. Sell. Sell.

'By 2006 the economy had picked back up again, and we had the best year ever. We'd picked up some extra data centres out of bankruptcy, sometimes at near-zero prices. We'd managed to get Google to put 100,000 servers spread across two of them. Business was good. One more push was needed. We kept up the energy levels, which was

more difficult than before, and returned to have another crack at floating the business.

We were expecting to float at £60-65 million, but it went down and down and down. In the end, literally eleven o'clock in the evening, the day before the float, our bankers, Investec said, "Look, we're only just covered. Cazenove's just phoned up and they said they'll take 35% of the deal, but at a 35-million-pound valuation."

It was a very difficult conversation, with Christophe practically walking out and me practically walking out thinking, "Well, how and why are we doing this?" We were absolutely knackered because we went into this process a second time, thinking that if it fails the second time we're really in trouble.

Now I was thinking, life's too short, we failed with one float, we can't fail a second float. The way the shareholder agreement was structured, at that valuation, the investors would get all the shares, with Christophe's and my equity reduced to a couple of percent each.

It was going to be a disaster, so I said to the investors, "Look, we'll do this, but I want a shitload of options. You're going to have to issue options within six weeks of the float, millions of options to make up for this thing."

Options are not as good because if you've got shares in a company that you've owned from the beginning, you'll just pay capital gains tax if you're lucky, you might even get some paper and then you pay just 10% tax. If you get given options, because it's like getting given a bonus, so you pay

45% tax on it, so there's a big, big difference in taxation, but hey, life goes on.

They shook hands and they said, "We can't write it in here. We can't give you a cast-iron guarantee, but we will shake hands on it." They were brilliant guys. Very, very trustworthy, very, very straight.

We were out there finally, but we were sub-scale. We were smaller than Telecity. We were smaller than Interxion. We were toying with the idea of trying to do something to reverse the merger – flattering ourselves that we could reverse into somebody else.

We knew that if we were sub-scale, we would just raise less money and therefore we would grow slower than everybody else. We'd shrink into insignificance as the years went by, while everyone else was consolidating and growing. We had to do something really dramatic, really full of energy, really transformational to the business to take it to the next step. But we were tired, and our investors could detect that. It was clear that Christophe and I just didn't have the energy to do that.

Suddenly, there was a little bit of a feeding frenzy going on. I found a good friend at Lazard and I said, "Look, we need a banker, can't pay you anything yet, but we do need a banker." We were in discussions about a merger with the CEO of Interxion at the time, Michel Broussard, that fell through, but now Savvis and Equinix got interested in buying us out and started fighting for our attention. AT&T joined in a week after and the race was on.

A great bit of work from Lazard and a few months later, in September 2007, we prepared for the integration of our business into Equinix. We had been a publicly quoted company for only 14 months, during which time our share price had gone up 6.4 times.

Over the last decade, I'd gone from working out of my bedroom to running 14 data centres across Europe, to floating on the London Stock Exchange, and eventually selling to Equinix. I was exhausted.'

Meeting the brand police

'Equinix is an American data centre operator, and with over 200 sites in 24 countries across five continents, they are the Big Daddy of the industry. They call their data centres International Business Exchanges (IBXs). Chairman Peter Van Camp compares them to "The Airports of the Internet". Even back in 2007, it was quite a monster – going around eating up smaller data centres quicker than Pac Man could eat up all the ghosts! We were one of those ghosts.

We jumped through hoops for Equinix and prepared a perfect integration of our business into theirs, but it wasn't all it was cracked up to be. In the early stages, Equinix suggested, 'Why don't we call you IX Europe, an Equinix company?' We thought, "Well, why don't you call us sec-ond-class citizens and you first-class?! Like 'Hi, I'm an Equinix guy and you are some little city schmuck!'" But we know that's the hold of Equinix because that's their brand.

And we had to complete an ear-out – so we had to do everything to make the company do well.

The day before the takeover we had a massive rebranding exercise. We had these funny little clippie poster frames in all the data centres, opened up ready to insert the new branding and we had T-shirts with Equinix on them, and then we had long discussions about how we could put all of the locations on the back of the T-shirts, a bit like a band on a world tour! All the technicians had black T-shirts with a list of sites which went alphabetically, so you'd have Chicago and then Duesseldorf and whatever and whatever, all the way through. It made us look as if we were all part of the same family very quickly. There was no divide.

So, by the morning of the changeover, everything had been rebranded. Our head of sales and marketing, Michael Winterson, was the head of the brand police. They were saying, "Is there anything of the old brand here? Remove it, change it!" Later, someone asked why we so efficiently removed all trace of our IXEurope brand. Michael answered, "because we love it so much". It was so successful that Equinix completely failed their next transaction because when they bought another company called Switch & Data in the US, they'd been lured into this false sense of security that the company they were buying was going to do all of this stuff that we did. They didn't. It was a complete disaster. The share price dropped 35%.

We'd made it too easy for them. We all just became Equinix. I became president of Europe for Equinix. A bunch of guys started turning up in Europe saying, "Hey, I'm Leroy from accounts" or "I'm, so-and-so, hey, listen why don't you do things our way?" I thought, "How about you go back to the US and stop bothering us unnecessarily?" I immediately instituted an expenses policy and said, "This is a really nice hotel at 75 pounds a night in London. That's a maximum budget and it's premium economy flights at best."

Everybody stopped coming over and asking a million questions, thank God, because that was a disaster. Still, they had a cast of thousands in the US, and we were a little tiny team in Europe. It was hard and, after a while, I remember we were trying to sign a deal in Paris on an extension for our data centre because we desperately needed more capacity, then suddenly the Equinix guys said, "Yeah, but I think we should do a feasibility analysis." I imagined writing a 300-page report and thought, "My God, I'm out of here. Maybe I'm not cut out to run big businesses."

On the 2nd June 2008, Christophe and I left Equinix. We felt we'd get in the way as strategy was now in the US and we'd become superfluous.

And that was it. At our insistence they had been very generous with share options for the remaining 14 senior staff, so the business was stable with a well-motivated team, and off we went. Off back to the bedrooms where it all began 10 years earlier. Off to start the next chapter of our lives, which, after a good one-year break, would end up taking me around the world to South Africa, then Russia and finally to Kenya, same old, same old data centre projects each time, with each project looking a little younger and smarter, and with some amazing teams of people involved.'

Guy's story is familiar to all of us in the industry who went through the boom and bust period of the late nineties to the mid-noughties. We were weathering a storm that never seemed to give in. We were all riding the waves the best we could, but just as we thought we'd hit calmer waters, the waves would come back higher than ever to pull us back under again.

And it didn't stop here. I'd survived some pretty serious storms over the last decade too – but the next one to hit would change my path forever.

CHAPTER 9

A QUESTION OF TIME

Tick. Tick. Tick.

Roll on to 2014, and business was booming once more. Power was what we sold to our customers, and so our size was measured in the amount of power we had available in our data centres to sell to customers. When we had floated in 2007, Telecity had 31 mega-watts of customer capacity, which was good by any data centre capacity standards and the equivalent consumption of around 10,000 homes, but by 2014 it had tripled to 97 mega-watts, with a construction plan underway that would take it up to 148 mega-watts within three years. This made us the biggest data centre operator in Europe, consuming as much power per day as a town the size of Canterbury in Kent, and it was just a matter of time before we were taken out by the Americans.

2014 was also a big year for me personally. On the 1st January, I was named in Her Majesty The Queen's New Year's Honours List. I had been awarded an OBE for my services to the digital economy. It was an incredible feeling finding out that I had been awarded the OBE and I was invited to attend an investiture ceremony at Buckingham Palace in March.

I was presented with the OBE by Prince Charles, who I'd actually met on lots of occasions because I was an active patron for The Prince's Trust and the British Asian Trust, which were both Prince Charles's charities. I'd spent a lot of time with the young people they support, helping and motivating them to get into work. So on the big day, when it was my turn to step forward and receive the medal, I bowed my head, and when I looked up, Prince Charles said, 'Oh no, not you again!'

I'd also published my first book by this time – *Forget Strategy, Get Results*, which was full of inspiring stories from my career and secured my label as a bit of a 'maverick' business leader. And in January 2014, it was my 50th birthday. Shalina threw me an amazing birthday party and she had a great big cake made to look like my book. It was in the Mayfair Hotel and she had arranged for ABC to do the music, with Martin Fry. It was an incredible night and memorable for many things, but especially one story I didn't find out until much later…

I was at Buckingham Palace on another occasion for a garden party, and I happened to be on my own. I saw Tony Hadley from Spandau Ballet, so I said 'hello' and he introduced me to this guy standing next to him. I said, 'Hi, I'm Mike Tobin,' and he says, 'Oh, yeah, I know who you are.' 'How did you know that?' I replied, 'Well, I'm also the manager for Martin Fry and ABC. Your wife, Shalina, was liaising with me and we were at your birthday party.'

It was quite a coincidence. Then he proceeded to tell me that behind the stage while ABC were waiting to come on to start the dancing, Shalina was getting ready to bring out this incredible cake that looked just like my book. Martin Fry was looking at it and he didn't realise that it was a fake book and put his finger in to turn the page, lifting up a whole pile of cake on his finger! His poor manager had to spend the next five minutes desperately trying to smooth down the edges of this cake to make it look like no one had stuck their finger in it.

So, all in all, it was a massively impactful time for me and I'd had an incredible first quarter of the year. I was being talked about a lot, but not everybody was happy about it, and I had a feeling that my professional future was about to change.

Trust your intuition

It was seven years since I had employed John Hughes as Non-Executive Chairman. We brought him aboard in 2007 to help with the IPO as – having been on the board of Thales, Lucent Technologies and HP – he had a good reputation for making these things happen. But I always had a niggling doubt about his intentions. On all of the boards he had been on as a non-exec, he had ended up moving into an executive role and then pushing people out of their roles and taking over. But he reassured me he had learnt his lesson and wouldn't be doing that again.

That promise didn't last long. Between 2007 and 2014, John gradually changed the shape of the board, pushing out members who didn't want to do things his way, and bringing in new people who had no previous affiliation with me or my team, and therefore no loyalty to me – only to John. He also gave very senior roles to people who should not yet have been at that level. Over time, I felt that the original members of the board who knew and who liked me were becoming marginalised. He was shifting the power and undermining my authority.

John was aiming, once again, to move from non-exec to executive chairman at Telecity, which would give him all the power in the business. Our relationship was getting more and more difficult to manage and we were both pulling in different directions. I was focused on our European growth, but he was also connected to Formula One and the Sultan of Oman, and he was looking at some way of getting the sport out there. He then decided to see if we could expand Telecity into Oman too. Remember, we were a European-based company and there was no way that it was logical for us to start stretching out to the Middle East at this time.

In 2014, John told me he wanted James Tyler, our head of marketing, to go out to Oman and see if running a data centre out there would be feasible. Of course, James was one of my direct reports and it wasn't for John to decide what he should be doing – but he'd gone over my head and started ordering James and others around anyway. I was not happy.

I said, 'Look, first of all, I think you shouldn't be setting unanimous direction for the company by sending out my guys on fact-finding missions and taking their time away from their daily jobs on a whim that you had in the Middle East, and also you didn't even tell me that you got two of my team involved already.' It was James and he sent Rob Coupland out as well.

Basically, they were told, 'Don't tell Mike. Just go and do this for me.' So this was really usurping my position. We had a big bust-up and that didn't help.

There was no possible good outcome anymore. I just knew this was going in one direction. It was just a question of time. We'd been getting into more and more conflict. He was even angry about my book. He said, 'Well, how can a public company CEO put out a book called *Forget Strategy*?' I said, 'Well, they can if they're in the 21st century and not the 19th century, and if you bothered to read the book then you'd understand.' So all of these things were building up to breaking point.

Just buy. Just sell.

John wasn't just shrewd at Telecity, he was the same everywhere he went. Over the past couple of years, I'd been helping Bo to recruit for Just Eat (people skills were not his forte). So I'd interviewed a new CEO for him, a Danish chap called Klaus Nyengaard, who was eventually replaced a few years later by David Buttress. But when it came to them preparing to IPO the company, they had no idea. I sat down with them and explained what it meant, but they needed someone to work with them on it, so at the time I recommended John Hughes. It made sense as they knew each other and John had the experience they needed to get them through it.

In March 2014, with John on the board, Just Eat was listed on the LSE's brand-new 'High Growth Index'. The *Financial Times* reported their movements, saying that they were the first company to list on this market, which was 'a rival to Nasdaq in the US but had taken over a year to get its first member[45]'. On the day they listed it, all the investor communities had built funds that could only invest in this high growth index. Just Eat was the first and only company on it, so all of them had no choice but to buy the Just Eat shares. As a result, Just Eat shares rocketed up on day one.

Then the very next day, John decided to pull them off of that market and put them on the main market, because, again, he wanted his non-exec chairman roles on the main market and not on the high growth market. But, of course, all of those companies who had bought into it in a big way (because it was the only stock on the index, and thus drove the share price up), suddenly had to sell all their shares in as quick a period because they were no longer there.

Consequently, the Just Eat share price plummeted, and so many people lost a ton of money simply because of the way that they decided to go about their IPO, which was all John Hughes's doing.

A few years later, I was told that David Buttress had suffered the same fate that I had endured. He left his role at Just Eat in 2019[46], and John Hughes moved swiftly into his position as Executive Chairman.

The end is nigh

Meanwhile, back at Telecity in 2014, John had made it a well-known fact that he always wanted to become a FTSE 100 chairman. We were on a very strong growth trajectory, and that basically meant that if we waited another two to three years we would break into the bottom of the FTSE 100. We were already in the very top part of the FTSE 250, so another two years of growth, and we'd be in the FTSE 100, we just needed to be patient. But John wasn't.

One of the things I had also been against over the years was to do a merger with Interxion. My problem was not the deal itself, but the valuation. Interxion and Telecity were very similarly sized businesses in revenue terms, but Interxion wanted to be valued at the same multiple as us, and we were actually a much better company in terms of profitability. We had much less debt, much more cash, and were a significantly more valuable company overall. It just wouldn't have been possible to do a 'like-for-like' share swap and merge them as equals. As listed companies, both businesses' share price were roughly the same multiple of EBITDA, but Telecity traded at a 50% discount on a PBT level.

We had discussed the deal with David Ruberg, the Interxion CEO, on many occasions and had never got anywhere. Ruberg also said that he didn't want to leave the company, instead, he wanted to carry on as CEO, and that would have created a big problem between him and me. So what this meant is that John getting rid of me would have effectively made way for him to make the Interxion deal. The combined business would then have a big enough market cap to get into the FTSE 100 and he would then have achieved his goal.

Waiting for the axe to fall

By August 2014, the atmosphere between John and I had become pretty bad, and I felt there was an inevitability to what was going to happen to me. But I was also pragmatic about it because I think that generally, things have a shelf life. Very rarely do CEOs last in public companies as long as I had, it was only because I'd built such a great business. It had gone from strength to strength, but I wasn't as buzzed and energized now because there wasn't a lot for me to do. The team was fantastic. They were doing a great job.

Had I been more financially stable at the time, I probably would have gone off and found something else. But I said to myself, 'Well, if I know what's coming, then I might as well wait for it and let them pay me to do this transaction rather than leave on my own accord and get nothing.' So I hung on and waited for the inevitable.

I don't know if John knew what I was thinking, or that I was on to him, but one day he pulled me aside to go for a coffee at Novotel on Kingsway. And he said, 'Well Mike, how are you feeling? Aren't you tired? Wouldn't you like to change your job? Wouldn't it be better?' And I said, 'No, I'm not interested.'

This went on for a while. And it just got more and more aggressive until one day, when I was due to be in Amsterdam, I was told to cancel my trip and come to the company secretary's offices in Hammersmith. We never had meetings in Hammersmith, so I knew exactly what was about to happen. I said, 'Let me know what the meeting's about so I can prepare.' John said, 'Oh, we want to discuss the future.' So I definitely knew what was coming!

When I got to the office in Hammersmith, John Hughes was there already, along with John O'Reilly, another Non-Exec and head of our Remuneration Committee. The meeting lasted about a minute and a half. I just walked in there and they read out a statement to me. My fate had been predetermined. The statement that the lawyers had prepared for me was a compromise agreement. It basically was a deal that would be announceable as both sides having mutually agreed to part company. And then John Hughes basically said, 'Look this is it.' I said, 'Well, we're not here to discuss, right? This is just you telling me.' And he said, 'Yeah.' And I said, 'Okay, fine then. Then that's it then.' And I walked out.

John had worked his magic once again. It was quite surreal. I had built this business up for the past 12 years from nothing and now I was being recycled! It felt harsh, but it wasn't a surprise.

The announcement about my impending departure was made on 26th August 2014. I had never really got on with the broking house Liberum over the years. They had always given Telecity a 'sell' recommendation and the analyst Joe Brent was continually found wanting in that advice[47]. But this time, they said it was likely the news of my departure would be well-received:

'We believe that Mike's departure will be well received as shareholders look for increased focus on return on capital and cash generation... We expect a positive share price [sic] reaction today...'

And then within a few hours...

'Shares fell 6.37% to 713p at 12:46 on Tuesday.'[48]

Part of the agreement was that I would stay on until the end of October 2014, which was when we put out our Q3 results. The business was still going incredibly well, and this allowed the market to see it as a planned process. So, basically, I was just a zombie in the office for three months; it was like being what Americans refer to as a 'Lame Duck President' when the new president has been voted for but the old one is still in office.

To the outside world, I still had the role, but I had no real power internally. I wasn't doing anything in the company, I was just working out the last three months and filling my day with looking for new opportunities. John didn't want me there, but he couldn't stop me coming in either, so I did.

We'd just moved into renovated offices in Canary Wharf. I had the corner office and I'd built in all this extra storage space around the walls because I had so much of my gear from the divorce that I wanted to store. So when I eventually moved out in October, I had to get a removal van to get all my gear out. It wasn't like you see on the TV where people leave with just a cardboard box with a family photo, and that's it. I literally had to have a removal van clear all my stuff out, because I'd been there for that long and was using it as a second home. It was a very strange feeling when I finally left.

The invisible man

I found out afterwards that, such was my legacy in the company, John had wanted to remove every reminder of me and make me invisible. He even turned my office into two meeting rooms, with a partition down the middle so that no one else could have it as an office.

He erased every memory of me effectively. And those of all the brilliant people that were part of that story too – Rupert Robson, Adam Soames, Richard Diffenthal – all the guys that had worked as advisors to the company, and people I am proud to call friends, who did such amazing work for us were literally just excommunicated overnight. He moved away from all the advisors that I had built up relationships with so that he could bring in his new advisors, ensuring there wouldn't be any kind of smell of Tobin left in the company.

That felt quite bad. But what amazes me is that out of all of the management team, which at its peak was about 15-strong, there is only a few left today working for Equinix (who, as I'll explain later, eventually bought Telecity).

All my team were very loyal to me. We kept talking, and they were really nice except for James Tyler. He had been one of my best men at my wedding but became quite bitter towards me at the end. It was very strange. I felt very let down, very betrayed by his change of approach. But I think he was a bit jealous. He did a lot of work around building my personal PR, and on the book, and I think maybe he felt resentment that he wasn't personally recognised in any way. But he's not really the sort of character that looks for that kind of recognition, he's a more reserved type of character, so I was shocked and disappointed to lose him as a friend. Perhaps he never was one, just a good actor. Everyone else was so very supportive, and were always on the phone. We stay in touch now. But he's the one that just went the other way. Ah well…

Stop right there!

By the time I left in October 2014, the plan regarding the Interxion deal had come out. Ruberg would be CEO of the combined business and John would be chairman, and it would just effectively pop into the FTSE 100 as a combined entity.

John Hughes had taken over my job as the interim CEO and a search for a permanent CEO was announced. But no one else was appointed and eventually he changed his title to Executive Chairman, on my full-time salary, just as he had done many, many times before in other organisations. He clearly had no plans to move back into a non-exec role. I was not surprised; it had been my gut feeling all along that John would eventually take my title.

On the 11th February 2015, the Telecity/Interxion merger was announced. Both share prices jumped up on the day and then, just as I had warned would happen, Telecity's went down again and Interxion stayed up, because the shareholders realised that an equal share merger was much better for Interxion's shareholders than it was for Telecity's. We were overpaying by swapping shares at the same value. Telecity would have represented 55% of the overall business and Interxion would have had 45%, simply due to the relative size of the businesses.

However, the Telecity story didn't stop there. Equinix CEO, Steve Smith, and his chairman, Peter Van Camp, had always said that they would never allow for a Telecity/Interxion merger as it would knock them off the number one spot in Europe and pose a threat to their position in the US too. So when they heard about the deal, they stepped in to stop it from happening by saying they would also make an offer for one of the businesses. There was a break clause in the Interxion/Telecity deal that meant that if a substantially higher offer was made for either party before the transaction concluded, that party could pay to exit the deal.

When they looked at the two businesses, Telecity was an infinitely better deal. It was the bigger business. It was the more profitable business. And now, because the share price had come down, it was also the cheaper business.

The 2015 first-quarter trading update for Telecity came out on the 7th May 2015, and still showed very strong growth and great performance. Alongside the update, there was also now an announcement that Telecity had received an approach from Equinix with an offer to buy the business.

The *Financial Times* reported their offer, which was a premium to the share price, offering to buy 'at £11.45 a share, a 27% premium on its previous closing price[49]'. The total offer was for £2.3 billion in cash and shares. And if they took it, John Hughes would move over to become a non-exec director of Equinix on a $1 million annual contract. It felt a little bit like he wasn't going to be working independently in his thinking about this and whether it was a good offer for the company. He was effectively getting $1 million to go ahead. Of course, John said yes, and Equinix announced that he would join the board following the close of the Telecity Group acquisition, which stretched out through the rest of 2015 to January 2016.

One of my former colleagues, who wishes to remain nameless but was working at Telecity, remembers the conflicts that were going on in people's minds at the time:

'There were these online tests that every employee had to do. In effect, it basically said that you weren't allowed to take a coffee from a vendor or a customer of this company without getting prior approval, literally. Everybody down to the janitor had to do this to show that we wouldn't take any gifts or bribes from anyone.

Just before the Equinix merger was announced, John sent a reminder to all employees how important it is that everybody does this anti-corruption questionnaire, how important it is that everybody fills this thing out, which was fine by us, we had done it.

What was just mind-blowing was that when the Equinix deal went through, there were significant compensations that were granted that were not based on any contractual stuff to John, despite the fact that he was acting CEO and the chairman of the board. He received a large pay-out for making this merger happen.

The juxtaposition of, on the one hand, insisting and talking about how important it is that you can't take a cup of coffee from vendors because, ooh, you might be corrupt and then, on the other hand, that happened. It was just something that a lot of people thought was just outrageous.'

Project Cheetah

By this time, it was Spring 2015. Since leaving Telecity, I'd been shopping around an idea for buying out Teraco in South Africa, the data centre operator that Tim Parsonson originally founded and Guy Willner helped develop. I didn't want to run it, just be on the board. I could see it was going to be big, and I tried four or five different private equity houses, trying to convince them to invest. None of them would.

I started working with Permira, the private equity guys (more about that adventure in the next chapter) as an advisor and they saw what a great opportunity it was and eventually bought Teraco in February of 2015 for around €100 million.

In the meantime, I had also spoken with Silver Lake and the Ontario Teachers' Pension Fund advising them to acquire Telecity and Interxion together, prior to their attempted merger because I knew they would be able to sell them onto Equinix for a profit, and I thought we could do a better deal with them than staying public. But none of that got off the ground despite many, many meetings all around the place.

So it all looked to be falling into place for Equinix, until the European Commission decided that, by acquiring Telecity, they were creating a monopolistic position on some of the European markets (London, Amsterdam and Frankfurt). The EC stepped in to put the blockers on the deal. The EC told Equinix that they had to divest eight data centres to another player in order to not be monopolistic in these four markets in Europe. These comprised of five in the UK, two in Amsterdam, and one in Frankfurt[50].

To overcome the issue, Equinix and Telecity carved out a little business from the combined entity and called it Project Cheetah. It was designed purely to facilitate the sale of the eight assets as required by the EC, and they treated it like a separate business running within the main business. This allowed them to get on with the process of integrating the rest of the Telecity acquisition, whilst divesting the required assets. The CEO of Project Cheetah had been appointed by Equinix and was none other than Rob Coupland, my original network engineer.

Project Cheetah was made up of the eight data centres that they were selling off, and now these were up for grabs to the highest bidder. We had a great opportunity to buy these assets, so I got Permira to bid. We just had to convince the European Union Commission that we were a solid data centre operating business capable of running them appropriately. We used the fact that we owned Teraco in South Africa, of course, as evidence that we knew how to run data centres and my intimate knowledge of those eight buildings because I'd either bought or built them during my time at Telecity.

The EC agreed, and we were successfully able to bid on the assets. Now for the negotiations.

On the Equinix team, there was a guy called Mark Adams, who was based in the US, and who we soon found out was extremely shrewd in his negotiation prowess.

The broker selling the Project Cheetah assets was The Royal Bank of Canada, represented by an American woman called Madonna Park and her team. She was based in New York and had always been very friendly whenever I was in the US with the Telecity crew because RBC wanted to get involved in our brokerage and try to work with the company.

When RBC got the gig to sell the assets, Madonna had no idea what she was selling. It was the first time RBC had ever worked with Equinix, and it was all new to her. She phoned me up and said, 'Mike, I'm coming over to London. Do you mind if we sit down with you and just try to understand what these assets are?' I said, 'Of course, no problem.'

So she and her team came over on a Sunday, and I spent the whole day with her in Rupert Robson's Torch offices as it was neutral territory for us, and don't forget I was working out of a home office now most of the time. We went through the whole thing so that she knew exactly what she was selling. By the time she walked into Equinix's London office on Monday morning, she knew exactly what she was doing, although she'd had no idea until the day before.

Later, I was in a meeting with Cheetah management at their lawyer's office with Rob Coupland and Martin Essig, who were both working for Equinix as result of the Telecity acquisition and had been assigned to Cheetah. We were discussing one of the assets in London that they were trying to sell. Now I knew these assets much better than anyone else because I had originally bought them, built them or competed against them when I was at Telecity. Equinix had no idea about what they had bought.

One of the assets was the Heathrow data centre, which is smack-bang in-between the current runway and the proposed new runway, so I knew that if the expansion runway ever went ahead, that data centre would effectively get wiped out when they knocked it all down. When I challenged them on the proposed Heathrow expansion construction and how that would affect the data centre, they said, 'Oh, no, no, no. It's far away, nowhere near the planned runway.' And I turned my computer around and showed them a satellite map of where the new runway was going to be and where their data centre was – right in the middle of it. This was the first time that anyone at Equinix had even realised that it was likely to be bulldozed!

So we had this much better knowledge of those assets than even the seller did, but we were bidding against Digital Realty, an American data centre operator who rivalled Equinix in size and market share, to acquire Project Cheetah. We got right through to the very last knockings of the deal when we were finally told that we had won, right at the end of the week. We were just waiting for confirmation from Mark Adams.

Then it all went quiet, and Mark said, 'Don't worry, we're just getting it ratified by a board meeting on the West Coast, which is later today, and soon as that's done you'll have confirmation that you've won.' I wasn't worried as I had faith that we were all working towards the same conclusion.

You should never trust bankers

We heard nothing. We waited through Friday, Saturday and Sunday, and then on Monday morning, we heard that Digital Realty had won.

Madonna Park had really done a number on us. RBC had told Digital Realty that we had offered up our Teraco data centre to Equinix as part of the deal, which wasn't true and wasn't even possible. The only reason we were able to do the deal in the first place was because we owned a data centre already – we couldn't possibly hand that over to Equinix.

So thinking that we had offered up a data centre, Digital Realty offered up their €200 million Paris data centre, which, at the time, Equinix leased from them. RBC basically said you can buy this data centre from us in return for doing the Cheetah deal. So she had played us off against each other at the last minute, even after telling us we had won it.

I only found this out later, over breakfast with Bill Stein, the CEO of Digital Realty, one of the largest data centre operators in the world. He told me what had happened. As part of Digital Realty's acquisition, Rob Coupland became CEO of Europe, the Middle East and Africa for Digital Realty, as well as MD for the Cheetah assets.

So that was all pretty painful but it's just one of those things. It also reminded me that you should never trust bankers. What goes around comes around though, and when RBC came to pitch for Pulsant's business (another company I am chairman of, and where Rob Coupland is CEO) they came bottom of the pile.

I must admit, however, the best Industry Analyst on the planet for data centres works for RBC and is a great friend of mine. His name is Jonathan Atkin. He knows what's going on in every data centre business around the world, not just at group level, but at country and subsidiary level. I try and catch up with Jon whenever I can, and we compare notes on the industry.

Mark Adams has now left Equinix and is running his own company, Adams and Associates, advising data centre companies. Good luck to him. He is a class act.

On Monday, 12th June 2017 the *Financial Times* announced that John Hughes had died after a short illness aged just 65[51]. He'd had a pretty eventful few years, but it was quite a shock. Shalina and I sent flowers to his family. He may have been a thorn in my side for a while, but all our experiences on this planet lead us somewhere, and it was time for us all to be thankful for our loved ones and move on.

PART 4

REACHING MATURITY:

Becoming a global advisor

BACK IN THE GAME

Man in the mirror

When I look back, leaving Telecity was quite a shock. I went from being in a mild state of panic in the first few weeks to being in overdrive for the next few. I used to have 150 emails a day and suddenly it went down to five. I'd been winning awards and featured in the media a lot, and now I was nowhere to be seen. I felt lost. When I looked in the mirror I didn't know myself. I'm also a narcissistic person, so I was thinking, 'Oh God, nobody loves me.' Even my kids were asking, 'What are you going to do now?' They could only ever remember me at Telecity.

I felt quite lost for a short period of time, wondering what on earth I was going to do next. I really didn't know what I was going to end up doing. I had a long think to myself and thought, well, to be honest, I'm too much of a maverick to get a FTSE 100

CEO role because they're all suited and booted posh people, and I didn't know a single FTSE 100 CEO that runs around with three buttons undone and ripped jeans! And I just didn't think I'd make it, and I couldn't afford to be anything less than that in a full-time job.

One day I was talking to Richard Holway, Chairman of TechMarketView. He said to me at the time, 'You know Mike, just take six months off, and reflect on what you want to do next.' And I remember thinking to myself, a) I can't afford to take six months off, and b) I only have a short shelf life. I'd just finished a role that was incredibly successful, and I had a great reputation of success. And in six months' time, I wouldn't... A lot of people forget stuff very quickly, and I didn't want to become one of those people they'd just forget.

So I went against the advice of my friends and some very wise people, but I felt quite scared of what was going to happen if I didn't. And I think that's how I treat fear now. I had to read my own book and talk about how I was feeling. I had to remind myself that when it comes to fear, either there's something you can do about it, in which case you can be afraid but get on and do it; or there's nothing you can do about it, in which case don't be afraid because there's nothing you can do! So I just got on and started trying to identify new opportunities, and my next role chose itself.

My plan was to build up a portfolio career quickly, but I knew that I needed an income on an immediate basis, because I needed to eat on an immediate basis. But I also needed to have some way of building a platform for the future and not having to work at some point in my life. So needing equity in things was an important element to the roles I took as well. So by working with private equity, I'd be a non-executive director, or I'd help companies build up to doing the transaction, and then I'd have an equity piece in the transaction, a non-exec salary. So I'd be earning a little bit a month on a salary basis but having the equity that would come to fruition in let's say three, four, five, six years' time gave me a longer-term plan.

Carving a new path

Within three months of leaving Telecity, I had four non-executive roles! My plan was working. I was still connected to the industry but now, instead of leading one company, I was advising investors on their global data centre deals and giving advice to the boards of the companies making the deals. I still had some very extreme highs and lows and would swing from euphoria when a deal was going well, to utter despair and not feeling like carrying on when it didn't. Sometimes, in the space of a few hours. It had been such a long and bumpy road to get where I had, it was going to take a lot of strength to get back in the driving seat and truly know where I was going, but I was starting to carve a new path.

One thing I was super sensitive about, having had that experience with John, was how management, executive management, in particular, treats non-executives on the board (or more accurately, fears non-executives on the board!) because they are worried that a non-exec, especially a chairman, will come along and start involving themselves too much or interfering with the business. And so I was extremely conscious of not wanting to be another John.

"

It was time for me to find that courage again

"

I made sure that I was positioning myself as a go-between for the executive management and the non-execs and the shareholders. So I was trying to sit in between all the stakeholders to try and give value to everyone, rather than just being a pain in the neck to executives. So that experience with John was a great lesson for me in what I'm doing now, because I think in life, it's very hard to learn lessons from successful events, because you don't actually know whether the success was luck or whether it was based on the key criteria. It's better, although perhaps mentally harder, to learn from the experiences that don't go to plan. It's like the high jump. The only time that you know when a high jumper has reached their maximum is when they fail because they hit the bar, that's it.

All I did know was that I didn't want to jump onto any old data centre company's board just to make up the numbers. I had so much real experience of the industry and I'd learnt from my mistakes and choices, so I wanted to use this to add real value. I also knew that, over the last decade, the European market for data centres had become saturated. So it was time to look at what deals could be had in the emerging markets, and one place that I had a lot of personal experience of in my younger years was Africa.

I'd lived in Southern Africa as a child during the apartheid era, and I'd escaped from the petrol bombs, bullets and an abusive father there, coming back to England with my mother when I was 12, only to be homeless. It had taken some real grit to get through it, but with a bit of luck and hard work it had paid off in the end.

At Telecity, we'd helped create the internet, and we did that by having a lot of luck but also having balls of steel too. It was time for me to find that courage again.

An African adventure

In 2009, after he'd left Equinix and had his own adventure in Russia, Guy Willner joined the founding board of Teraco Data Centre in Johannesburg, South Africa. Originally founded by Tim Parsonson, it was now run by Lex van Wyk as CEO and Jan Hnizdo as CFO. Nowadays, I count them both amongst my very good friends, and Jan has recently been promoted to CEO of Teraco. But back in 2009, South Africa was a developing nation where jobs were growing and the internet was just starting to take hold. People were getting increasingly impatient at seeing the 'circle of doom' spinning on their computers as they tried to download content that was being pulled across from the US and Europe with terrible latency.

South Africa was able to provide a data exchange service that was cheap and efficient for all the big players. Teraco is unique in the industry, in that it is based inland in Johannesburg, a city that has grown up around the gold mining industry. All the big banks and the biggest stock exchange on the African continent are located in Joburg, where, as the mining industry started to unwind, industry took over and it became the financial hub of SA.

Guy had seen the potential for South Africa to have its own vendor-neutral data centre where all available carriers could be reached in one go, rather than split up around multiple sites.

Teraco had, over a few short years, become one of the major drivers of the costs of connectivity coming down in Africa because it wasn't just the big telco providers that could afford to connect via Teraco, the small challengers could too – reducing the price of connectivity by up to 40 times! Teraco was also the catalyst for 'dark fibre' coming to the region because they could link up to all the ISPs and telcos in one place.

The company was initially funded by venture capital and then, by 2014, the investors, Treacle, Pentangle and Marlow Capital, had been in for five years, and a VC typically wants to exit after that sort of period[52]. The company, led at the time by Tim Parsonson, decided to go and find a new shareholder.

At the same time, I had seen the potential for growth in SA and I had already approached Permira about Teraco and was in talks with them to bid for it. I remember having a final meeting at the Connaught in Mount Street with Richard Sanders and Michail Zekkos. We had a load of cocktails and ended up getting very drunk. In my efforts to convince them to close this deal in South Africa, I wrote on a cloth napkin, 'Teraco will be a 10 bagger within five years[53].'

The funny thing is they took that and had it pinned up on the wall inside Michail's office for the next four years and, of course, four years later they sold it, in Rand terms, for around 14 times more money, which was an amazing return. Obviously, there was some currency depreciation, but Rand for Rand it was a 14x return on investment.

When it came to making the deal, Teraco were not particularly interested in Permira investing as they didn't seem to have an interest in the industry or Africa. That's where I came in. I was able to give them the technical expertise to make sure they were investing in a credible business and to give Teraco the reassurance that, with my industry experience on the team, they would be able to grow at pace going forward.

We went to visit the Teraco data centre and, to everyone's surprise, the first thing I did was get in and start lifting up floor tiles. They asked me what I was doing and I said, 'I want to see what your wiring is like,' because if there's one clear sign that a data centre is well managed, it's what it looks like under the floor! If the cabling is tidy and it's not too dusty, it's a very good sign. It was very tidy under there and I knew this was a very well-run facility!

Funnily enough, I just did that again recently at another data centre in Europe and they couldn't even find the suction cups that you need to raise the floor! Eventually, they managed to get a floor tile up and then I understood why; the cabling underneath was a complete shitstorm.

It's a coup

In January 2015, on the eve of closing the deal, I was in Johannesburg signing all the documents in the lawyer's offices, and then I left. The Teraco team were closing the deal later that evening so they signed their docs and were due to fly to London later that day. As luck would have it, I had a box at Twickenham stadium and the next day, Saturday, England were playing rugby against South Africa. Now South Africans are crazy on rugby and Lex and Jan were no different.

On Saturday morning, the guys were supposed to meet us in the box at Twickenham for lunch before the game, but by 12 noon they still hadn't arrived. I called them and could detect they were in the UK as there was a UK dialling tone. But they weren't picking up...

Eventually, we got through and Lex explained that as they got onto the plane in SA to come to London, having signed all the documents and left them with the lawyers, Equinix had flown a massive team into Joburg, and stormed in with a 50% premium to our offer to try and usurp the deal at the last minute.

It was a proper boardroom coup. Rather than coming to the match, the team had been on the phones since they had landed trying to get to grips with the rest of the shareholders, who had, of course, wanted the higher offer to win the deal!

Management told me they wanted to do the deal with us but didn't want to upset the other shareholders by refusing the higher offer. I told them to say this to their shareholders: 'We are not going to stop you guys doing this transaction with Equinix if you want to, but we don't want to be part of that story going forward and will resign.' And I said, 'I guarantee that Equinix will not do the deal because they couldn't take the risk of a big investment, their FIRST investment in Africa, without management coming along for the ride.'

So it was a big gamble for the Teraco management team because I could have got it wrong, and they may have ended up quitting the company they had built for nothing! But fortunately, it worked out well. Teraco weren't ready to be corporatised, as Jan Hnizdo explains:

MICHAIL ZEKKOS AND JAN HNIZDO ON A SHEER FACE OF
TABLE MOUNTAIN, CAPE TOWN. A CHALLENGING BOARD MEETING!

'We really wanted to do a good thing for the internet in South Africa. The business was growing and we loved the brand, so you can imagine you've got a whole bunch of passionate people that have grown this company and who felt it was way too early to be corporatised into someone else's business. And that's basically how it went down in the board meeting, we said, "Look, you know, we all love the business, we all love the brand, it's way too early to do it and we'd prefer to do the private equity option as opposed to the trade buyer."'

So the deal went ahead and I joined the Teraco board as an investor and a non-exec director. It was time for my South African initiation.

A game of fives

Our motto on the Teraco board was to grow the business but have fun while we were doing it. And we were lucky to have a really interesting board with an eclectic mix of characters, including Lex Van Wyke, Jan Hnizdo, Andrew Tuckey, who is a proper Rhodesian gentleman, Michail Zekkos, and a very bright kitesurfing lad, Pierre Pozzo from Permira.

When I met Jan Hnizdo, Teraco's then finance director (he has subsequently been made CEO), he decided to test my steel by introducing me to a great South African custom called fives.

'We have a custom at Teraco to get to know people, we call it a game of fives. Which is a drinking game based on little skill, and lots of luck.

Everyone puts their hand in a circle on the count of three, and you have the choice to put your closed fist in, representing the number zero, or an open hand representing the number five. You then take turns to guess how many fingers in total will be shown each round. Guess right and you are out. The poor guy left at the end lands up having to have a penalty, which inevitably is three shots of tequila or whatever horrible drink that we've got in the bar that we can't seem to get rid of.'

Well, as a non-South African, clearly, I don't have the expertise and skills dating back many years to win this game, so I tended to lose quite often. But then again, I also think it's probably because of the fact I was quite happy to have the drink(s). Whenever we had board lunches, we would inevitably start playing fives – we're probably the only board in the world who did!

Once we had a board meeting in Cape Town booked for nine o'clock in the morning. But the previous night, we were playing fives and we thought it would be a good idea to go to Table Mountain at five o'clock in the morning before sunrise. Jan recalls the story:

'There's an easy route up the mountain, there's a hard route, and then there's the other route. The one only I know, apparently. And we thought, "It'll be fine, we'll quickly get up the mountain in time to take the first cable car that comes down."

So it's still dark and Mike, Pierre, Michail and myself are planning on getting to the top of the mountain, getting the first cable car down, getting back into the hire car, and

"

Whenever we had board lunches, we would inevitably start playing fives

"

getting back to make our nine o'clock meeting with the board and a senior Permira partner who was dialling in to join us. We figured that with all these marathons Mike had been doing he'd obviously be really fit and would lead us around. But he was still drunk from the game of fives!

The hike proceeded to go quite well and then at some point halfway up the mountain there were some signs showing the easy route, the hard route and the even more difficult route. Anyway, I convinced the guys to take the harder route, because the harder route is more enjoyable.

So like mountain goats, off we go. But somehow we landed up going completely off course and there was no route. Within 20 minutes, we were basically scaling cliffs. Table Mountain's got these cliffs and then there's a bit of under-growth under each one. So we'd landed up walking along these cliffs and I tell you, by the time we finished the walk, I was five minutes away from phoning mountain rescue. Because it got to the point where we were on a ledge and basically, one of us was going to fall off and die. And it was probably going to be Mike because he was the most pissed.

The two guys from Permira were so concerned, not about themselves, but about the fact that we were going to end up killing Mike. And we had several debates saying, "Do we phone mountain rescue or do we carry on?" We ended up pressing ahead and to keep everyone going I kept saying, "It's around the corner, it's another five minutes. It's another five minutes. It's five more minutes, five more minutes." It took four hours.

Our board call was supposed to start at nine in the morning but landed up starting at about one o'clock because half the board were stuck up the mountain.

Then, in the car heading back, I tried to get us to the meeting as fast as possible. I drove down all these winding roads and we had to stop on the way back for Mike to throw up!

I don't know whether it was purely because he was hungover, or whether he was in shock, but that drive back was probably as dangerous as being up the mountain on the cliff edge!'

We don't do things by halves! We had a number of AfricaCom events around this time too, which is an industry conference traditionally held every year in Cape Town when all the African telcos and the European players throw parties. There was a particularly memorable Seacom party, which is in Granger Bay, literally on the beach. Seacom is a major telecommunications company in the region and a main sponsor of the conference.

Jan Hnizdo recalled this typical night out:

'We started the evening playing fives, as we normally do, and somewhere along the line, at around about ten o'clock, none of us could seem to be able to find Mike, so we all assumed that he had taken a taxi home and settled for an early night.

At about twelve o'clock the party was kind of, you know, coming to an end. And for whatever reason, we went and searched the loos. We found Mike passed out in one of the cubicles. Not a good idea! It's incredibly dangerous passing out in a public toilet somewhere in South Africa!'

Needless to say, I've learnt a lot from the South Africans already! But, hopefully, I've taught them a lot too. We sold Teraco in February 2019 and my South African journey was over. Just four years after acquiring it for well over €1 billion and the business had become the 'game-over' leading data centre operator for the entire African continent. They had grown from around 6MW to over 50MW of power and become a leading name in the industry. Both Lex and Jan are still there, although Lex has moved into a more statesmanlike role and Jan to CEO.

I am so pleased for these guys who I am proud to call friends. Jan and I have a tradition of betting on rugby and cricket matches around the world and, in 2019, I was in Japan for the Rugby World Cup final, which fittingly was between England and South Africa. But as England lost that match we won't talk about it!

Head in the clouds

Over the last 10 years, all of our personal and professional needs and expectations have changed when it comes to storing and accessing data. In response, data centres have moved from being based purely on rack space for servers and hardware, with traditional infrastructure, to utilising the 'always on, never fails' benefits of pay-as-you-go cloud storage.

Using the cloud supports greater scalability and it doesn't need to be one or the other – data centres now provide both cloud storage and dedicated options without negatively impacting on the other. So over the last decade, the industry has grown to include both pure-play data centres who focus exclusively on servicing their customers' data needs via traditional hardware servers on racks; cloud-based data centres whose data is stored remotely in the cloud; and hybrid providers who offer both services.

In the years since Telecity, I realised that my long journey through the ups and downs of the industry gave me a real strength in being able to give trusted, credible and highly valuable advice to the new kids on the block. I was able to take different kinds of cookie-cutter templates of how to do business in this industry and be best in class, and use them to create new successful data centre companies. This made me an attractive proposition to work with for financiers looking to purchase data centres, as well as the companies selling them. The balls of steel were coming back!

Spanish inquisition

A good example of how traditional data centres have evolved over the last few years can be seen in the deals I got involved in shortly after my exit from Telecity in 2014.

In 2015 I was aware of a couple of data centres in Spain that we had looked at for Telecity's expansion a few years before, but nothing had ever come of it. They were owned by Schneider Electric and were being spun off as an entity called Telvent. Javier de la Cuerda was the CEO at the time.

Harro Beusker from Equinix flew down and had a look at the asset but turned it down. But I knew a great Spanish guy called Fernando Chueca, who was based in London and working at Carlyle. He was looking at assets in Spain because he believed the economy there was about to take a turn for the better, and being Spanish himself, he had a strong value-add when it came to working with them.

Meanwhile, Telvent was losing money and Schneider wanted to sell the assets for around €34 million, which inspired Fernando to make a move. He thought, 'Data centres are pretty popular at the moment, and that's pretty cheap, there must be some value in this.' So he called me up and said, 'Do you want to help me try and unlock some value here?' I said, 'Yeah, sure.' So, in 2015, he bought the business for around €34 million.

Next, we bought another business called CloudMas, 'cloud' as in the internet, and 'mas', meaning 'more' in Spanish, so 'more cloud'. It was a six-man business, and we spent a few million on that, bolted CloudMas and Telvent together and built a strategy on the back of creating a new hybrid cloud solution. There was a great CEO at CloudMas called Fernando Negra. We parted company with Javier de la Cueda and brought in Faustino Jiminez to see if we could turn the business around.

At the time, I was also on the board of Datapipe, a data centre operator founded by Robb Allen in the US, which was an Abry Partners private equity asset. They had asked me on the board because they wanted somebody to represent Europe for Datapipe. That company eventually rolled into Rackspace, which is one of the larger hybrid cloud providers. But Datapipe was revolutionary due to its positioning in the hybrid cloud space, so I had all the techniques to help reposition Telvent to the market as a hybrid cloud solutions provider.

When it came to Telvent, I knew exactly how to get the best value out of it for Fernando. First, we renamed it Itconic, completely rebranded it, and then made it into a hybrid cloud solutions provider and carrier-neutral data centre company, which is where the industry was heading. About 25 months later, we sold it to Equinix for around €215 million. The little business they had turned down at €34 million just a couple of years earlier!

The actual process of selling it was quite funny. Equinix wanted it but Harro Beusker was like, 'Okay, well I'm not paying more than 80 million Euros for it' and, of course, Carlyle were like, 'Yeah, that's great, that's fantastic, because we spent 30 million and we've got 80 million just two years later.'

I went nuts and I said, 'Are you kidding me? This is ridiculous. This is completely undervaluing what it's worth to them.' This was the last independent asset in Spain. Interxion was in Spain, but Digital Realty and Equinix were not, and this was the last meaningful independent asset that they could get without having to go to greenfield and start from scratch.

So I suggested we start some competitive tension between them, which we did, and we used Torch Partners again to help us. We played them off against each other a number of times, and then Harro came on the phone and said, 'Okay, I'm offering 130 million, and that's it'. So Fernando accepted his price, and again I went nuts and said, 'You're really undervaluing this.' He said, 'But I've given my word. I can't go back on it now.' I said, 'Do you think Harro cares about your word? I've been ripped off by Harro with the Cheetah assets just recently and many times before. He doesn't care about his word, so go back to him and say, "It's got to have a two in front of it or there's no deal".'

So he went back to Beusker and told him what we'd agreed, and Beusker said, 'I'm never speaking to you again.' And slammed the phone down on him.

Fernando came back to me at Carlyle and said, 'Mike, you just lost me a ton of money because we were standing to make five times more money there and the guy has just walked away.'

I said, 'I promise you he has not walked away.'

He said, 'Yeah, but he hung up and swore to me he'd never speak to me again.'

I said, 'Trust me, he'll be back.'

Mark Adams of Equinix rang Fernando two days later and said, 'I'm really sorry, it's 200 million.' They had come back with the minimum 200 as predicted. Again Fernando said, 'Yeah, that's fine,' and I said, 'No! Just because I said it needed to start with two, doesn't mean to say it would finish with zeros.'

So we carried on negotiations and eventually, the price got to €215 million and in October 2017 Equinix bought the business. Recently, at a conference, they said it was one of the greatest deals that they had ever done, as it 'strengthens Equinix's position in Europe'[54] which shows the relative value propositions to different people.

Dealing in data

Today, around 40% of the world's population has internet access. Data centre brands have matured, merged, collaborated, and acquired each other in a bid to be the biggest and best provider. The nineties and noughties heydays of champagne lifestyles and private jets have now evolved into a highly respectable and valuable future-proof industry. The whole world has moved forward and more and more regions are now emerging as viable environments in the data centre space.

In the last five years, constant mergers and acquisitions have developed the industry and changed the landscape. As more capacity came in from systems integrators who also owned data centres, some of the traditional data centre operators, such as Equinix, started looking into managed services as an enabler to win customers for their main product, which is colocation, or standard space and power.

In 2016, Host Europe was sold to GoDaddy. GoDaddy spun out a bit of it called PlusServer, which provided hosting to enterprise clients. I had advised the private equity house BC Partners to buy PlusServer and use it as a platform to consolidate the managed service industry across Europe, but that went sour. Firstly, I helped them get into a business called NextInto in Hamburg, where my good friend Dirck Hanssen was sales director. BC did the transaction, but then promptly fired my friend without even telling me. This really upset me and started to erode the trust between me and the company.

Shortly after, Basefarm in Norway, a managed service and data centre company based in Oslo, went up for sale. I was on the board of the asset on behalf of the owners Abry Partners. Abry is a great team that has made a lot of great investments in the sector and this was no exception.

Basefarm was an ideal acquisition for PlusServer, allowing them to strengthen their position on the German market and have exposure into the Netherlands and the Nordics. I was on both the buy-side and sell-side (in other words, I was acting as a director/advisor of both the seller and the potential buyer at the same time), so I backed out of the negotiations after making the introductions. In the end, after using up a lot of management time, PlusServer chose NOT to bid on the asset, which again was a bit embarrassing for me, having only found out after the event. However, my good friend Rupert Robson had been acting for Orange Business Services and swooped in at the 11th hour and acquired the asset.

It was clear that BC Partners intentions around PlusServer were no longer pan-European, especially when they appointed a new CEO, Oliver Maus, and I only learnt via an announcement on LinkedIn (in German)! PlusServer and I parted ways.

In 2015, I was approached by Oliver Jones and Jonathan Berney to get involved in a Hong Kong-based Chinese data centre operator called Chayora as an investor and a Non-Exec Director. Alongside Africa as an emerging data centre opportunity, Asia was also the stand-out region for pioneering assets in the data centre space.

Initial investment came from Standard Chartered Bank, but there was a hiatus in terms of construction following a management buyout of that division. The MBO was swallowed up by Actis, a global growth markets investor, and in 2019 further investment was finally secured to build the first 13-megawatt phase of a 300-megawatt facility in Tianjin near Beijing. An interesting lesson in how long it can take to get capital for the industry when it is in an emerging market! During this time, I also managed to convince Actis to make an investment in Nigeria through the acquisition of Rack Centre in Lagos.

I was also on the board of IXCellerate in Russia with Guy Willner over the same period of time. Another great asset that, once the political tensions between Russia and the USA subside, will be sold to the big boys for what we expect to be a market-leading position on the Moscow marketplace.

All these data centre assets will come into their own over the coming years, and eventually, the industry giants will have to acquire them, and will do so at a premium as they can move faster, rather than trying to attack those markets on a greenfield basis, buying up land to build data centres from scratch. In addition, building connectivity rich ecosystems can take decades organically even when done well.

The law of big numbers will force the real estate investment trusts (REITs) to look further afield than the mature Western markets in the US and Europe; instead, they will look to grow into new markets where they can get first-mover advantage with the premium asset. They will also look into regional data centres outside of the core capitals in mature marketplaces. The likes of NorthC in the Netherlands and Pulsant in the UK are great examples of data centre operators that have built portfolios of assets outside of the capital markets in their countries and focused on the regional play.

Back to the future

When I found myself in the data centre industry back in 2002, I could never have imagined the stories, opportunities and relationships it would bring me. Neither could I have imagined how fundamental the industry would become to everybody's lives across the planet.

Thanks to data centres, today we are more connected than we ever dreamed possible. We have more opportunities at our fingertips than we could have ever imagined. And every day we have more knowledge, skills, experiences, interactions, activities and accomplishments than ever before. But it doesn't end here. This is just the beginning.

Back in the day, building value for data centres meant buying up more and more retail space to put your servers in. It was all about the physical space. And today that's still important, but on top of space, we need power and connectivity and the infrastructure to keep growing our means of connecting regions, countries and continents. How do we do this? It's all about the subsea cables. Vinay Nagpal, the founder of InterGlobix, a global consulting and advisory firm focused on the convergence of data centres, subsea and terrestrial fibre, explains:

> 'The key value proposition of subsea cables is that they connect, just like a data centre, connectivity is important to connect it to the rest of the world; ultimately, you need to have robust, low latency, high bandwidth connectivity between one part of the world and another. Subsea cables connect different countries and continents together.

Microsoft and Facebook drove the first subsea cable to land from Bilbao, Spain, to Virginia Beach. It was called the Marea cable, it was the fastest subsea cable to be put into existence in the history of subsea and in under two years, it was already at maximum capacity.

Telxius, the Spanish telecommunications infrastructure company, got involved in operating the cable. They had a second one come from Rio de Janeiro in Brazil with the branching into Fortaleza and Puerto Rico in San Juan, ultimately coming to Virginia Beach. Now Google is bringing a third one from France to Virginia Beach.

So the future is about having subsea cables which are landing more and more into a connectivity rich, carrier-neutral, cloud-agnostic data centre versus just a cable hut by the beach.

It took 161 years to have the 448 cables that exist today, covering around 1.2 million kilometres in total distance. But in just four years, from 2016-2020, we created 107 new cable systems, representing 400,000 kilometres, and roughly around $14 billion of value. And a lot of these new systems are driven by the "new age" subsea fibre operators and the hyper-scalers, including Google, Facebook, Amazon, Microsoft. Wherever they're going with their new cloud availability zones, they want to have more management of and control over the network component, hence their investment in subsea. Subsea cables are becoming the arteries of the internet.'

What Vinay really does articulate here is what's going on in terms of the infrastructure behind the internet. We talk about connected cars, the Internet of Things, big data, home networking, connected cities, smart cities, smart homes etc. etc. But we all take the infrastructure behind the internet for granted. And we all have our moments when we moan that we can't download this, or watch that on our TVs, but actually, there's billions and billions of dollars being spent in putting infrastructure above and below the ground and under the sea to make all this possible every single day.

We just look at our remotes and want to have access to Netflix, Amazon Prime, and all that content, instantly, but there is underlying infrastructure that needs to be upgraded. And there's new infrastructure being added to support not only 4K, but 8K and 12K video broadcast over the internet.

Now look at home networking. In new houses, every single thing that can possibly be linked to the Wi-Fi is IP enabled, from your laptop to your lightbulbs, TVs to thermostats, toasters, fridges, blinds, doorbells, garage doors, cars and so much more. And then there are all your family's devices… which, by the way, the analysts say will put the total number of devices connected to the web to over 50 billion around the world by 2030 [55].

We talk to Alexa, play music, order food, check the weather, plan our holidays, work, rest and play. All thanks to the internet. And all thanks to the industries supporting it 24/7.

When I counted, I found I had around *280 IP addresses* connected in my home. The fact that I have worked in the data centre industry for the last three decades has no relevance to how connected I am today, or how much more connected I will be in the future. The internet does not discriminate between those who know how to lift the floor and get under the tiles, and those who don't. It gives those who can access it all the same opportunities. This is the story of how those opportunities have been brought to your door... what you do with them is up to you!

AFTERWORD

As the Internet of Things becomes more prolific, localised capacity (by which I mean the proximity of the processing power to the end user) will be more and more important, and ultimately we will see vehicles being mobile data centres and edge or micro edge data centres on each street corner. But between now and that time, the first move will be to go into the regions and secondary towns and cities.

The major data centre operators are turning to these regional markets and more exotic emerging markets to continue their growth and thus their shareholder value. They are also looking for differentiation from their competitors.

Investment is coming from new providers, in particular the infrastructure funds, who have become more comfortable with the asset class in recent years. They have lower returns profiles compared with traditional Private Equity, and this is driving valuations up for the right assets.

Data centres are a unique asset class. They are essentially glorified real estate, with a combination of relatively finite attributes, including large amounts of power on a specific location, with planning permission and connectivity, but on the other hand, the demand for them is governed by the continuous adoption and use of the internet. Just like dark fibre, subsea cables and mobile towers they are benefitting from a new data revolution. We cannot imagine using the internet less tomorrow than we do today, and we are yet to see the full impact of Big Data, the Internet of Things and AI, and Machine Learning on global data volumes. Therefore, it is very reasonable to say that these assets can only continue to become more and more valuable, and more and more important in our lives going forward.

Sadly, the world has experienced the life-changing implications of the global COVID-19 pandemic. Almost overnight, our liberties were curtailed, travel was scrapped, and the hospitality industry was decimated. But the internet continued to boom, with the likes of Zoom and other video conference systems seeing incredible take-up. To 'Zoom', like Google and Uber before it, has become a verb. Supermarket home delivery systems have crashed through increased demand and restaurants' websites have become online delivery platforms. Virtual dating, online training courses and social media have gone mad. Everything we are doing can only happen because of the data centres and often hidden resilient infrastructure around the world.

Massive gains will be available for those with the courage to enter into emerging markets ahead of the competition and create market-leading positions in those geographies. The art of calculating leading edge versus *bleeding edge* (and yes, it is indeed an art) is critical in terms of cash-flow management, but those who have delayed entering the industry over the past 20 years have generally regretted that decision.

Data centres are as intrinsic to modern life as airports, and are getting more important every day. I am privileged to have been part of the creation of this great industry which, in 100 years, will be looked back upon with the same importance and reverence as the railways in the Industrial Revolution.

ACKNOWLEDGEMENTS

With sincere thanks:

To Shalina, Eloise, Nelson and Rose for their unending patience in me.

To all the industry teams I have worked with, who will forever be part of my story. I wouldn't be here without you, in one way or another. Special thanks also to those who have been so generous with their own stories and memories for this book, including:

Stephane Duproz for fighting fires, of all kinds.

Martin Essig for his enduring support and friendship.

Guy Willner for his honesty, energy and inspiration. Josh Joshi for his openness, calmness and perspective. Vinay Nagpal for his expertise and encouragement.

Milan Radia for his incredible memory and industry insights.

Jan Hnizdo for keeping me alive up mountains (just), and beyond.

Joe Valenti for longevity of loyalty, regardless of time and distance.

Philip Low for recognising and rewarding my resilience.

Adam Soames and Richard Diffenthal for keeping me on the right path.

Rupert Robson and the Torch team for countless amazing successful deals.

All my Telecity Managers (except James) and those I have worked with over the years.

To Donna O'Toole, Leila Green, and Ali Dewji for their creativity and clarity.

ENDNOTES

1 https://www.thetimes.co.uk/article/telecity-shows-its-maverick-michael-tobin-the-door-5k8205k33wf

2 https://www.euractiv.com/section/digital/news/commission-internet-under-strain-amid-covid-19-overuse/

3 https://www.weforum.org/agenda/2017/09/the-value-of-data/

4 https://www.dailymail.co.uk/news/article-4558034/Was-human-error-blame-BA-chaos.html

5 https://www.reuters.com/article/us-uber-ipo/uber-unveils-ipo-with-warning-it-may-never-make-a-profit-idUSKCN1RN2SK

6 https://www.wsj.com/articles/SB1007092967508235920

7 https://en.wikipedia.org/wiki/Commodore_PET

8 http://savethesounds.info/

9 Nothing Like a Dame: The Scandals of Dame Shirley Porter, Andrew Hosken (Granta Books 2006).

10 https://www.independent.co.uk/news/business/news/stanford-loses-fight-to-oust-redbus-executives-172302.html

11 https://www.theguardian.com/technology/2001/dec/22/internetnews.business

12 https://www.prweek.com/article/98961/diary-harvard-centro-revs-redbus-global-trip

13 https://www.theguardian.com/technology/2001/dec/22/internetnews.business

14 https://www.theguardian.com/technology/2001/dec/22/internetnews.business

15 https://www.telegraph.co.uk/finance/2861599/Redbus-sues-Stanford-over-Malaga-villa.html

16 https://www.questia.com/library/journal/1G1-89491839/charlton-athletic-moves-into-the-fast-lane-with-redbus

17 https://www.zdnet.com/article/cliff-stanford-the-maverick-internet-pioneer/

18 https://www.independent.co.uk/news/business/analysis-and-features/stanford-fighting-to-take-control-of-redbus-186571.html

19 https://www.independent.co.uk/news/business/analysis-and-features/stanford-fighting-to-take-control-of-redbus-186571.html

20 https://en.wikipedia.org/wiki/Homes_for_votes_scandal

21 https://www.theguardian.com/housing-network/2012/feb/01/fred-goodwin-scandal-dame-shirley-porter

22 https://www.independent.co.uk/news/business/analysis-and-features/stanford-fighting-to-take-control-of-redbus-186571.html

23 https://www.investegate.co.uk/article.aspx?id=200207300700212667Z

24 https://www.investegate.co.uk/i-spire-plc--is--/rns/final-results/200212191037053451F/

25 https://www.theguardian.com/business/2002/oct/18/3

26 https://www.forbes.com/sites/rachelsandler/2019/09/10/cannabis-king-boris-jordan-chairman-of-curaleaf-becomes-the-only-pot-billionaire/#1b79517f37dc

27 https://www.telegraph.co.uk/finance/2857307/Redbus-founder-fails-to-oust-board.html

28 https://www.independent.co.uk/news/world/europe/reward-raised-to-find-au-pairs-killer-8813559.html

29 https://www.telegraph.co.uk/expat/expatfeedback/4198173/36-years-for-kidnapper-who-stripped-and-strangled-fiesta-girl.html

30 Nothing Like A Dame, by Andrew Hosken (Granta Books 2006)

31 www.ii.co.uk

32 https://www.independent.co.uk/news/uk/this-britain/sleaze-scandal-strips-dame-shirley-porter-of-her-title-94961.html

33 https://www.telegraph.co.uk/finance/2876779/Internet-tycoon-charged-with-blackmail.html

34 https://www.telegraph.co.uk/finance/2876779/Internet-tycoon-charged-with-blackmail.html

35 https://www.theguardian.com/technology/2005/sep/16/egovernment.politics

36 https://www.theguardian.com/technology/2005/sep/15/guardianweeklytechnologysection.uknews

37 https://www.consumeractiongroup.co.uk/topic/335636-ifonic-claims-ltd-swansea-86yrs-old-apparently-%C2%A3990-out-of-pocket/

38 https://www.cnet.com/pictures/shame-shame-decades-10-biggest-tech-scandals-images/2/

39 https://www.techrepublic.com/blog/10-things/the-10-biggest-tech-scandals-of-the-decade/

40 https://money.cnn.com/2006/03/08/news/newsmakers/scores.suit/

41 https://en.wikipedia.org/wiki/Hustlers_(film)

42 https://pressreleases.responsesource.com/news/16643/telecity-appoints-josh-joshi-as-group-finance-director/

43 https://www.3i.com/media-centre/corporate-and-portfolio-news/2007/3i-backed-Telecity-group-achieves-successful-436m-ipo/

44 https://www.ft.com/content/afffb542-f577-11e9-b018-3ef8794b17c6

45 https://www.ft.com/content/47adfc72-b4d9-11e3-9166-00144feabdc0

46 https://www.ft.com/content/86acbac-ef61-11e6-930f-061b01e23655

47 https://www.ft.com/content/715a7fb0-c188-11e0-acb3-00144feabdc0
https://www.sharesmagazine.co.uk/news/shares/small-cap-opportunities

48 https://www.sharecast.com/news/broker-recommendations/telecity-a-sell-on-ceo-departure-says-liberum--589443.html

49 https://www.ft.com/content/85d50414-f493-11e4-8a42-00144feab7de

50 https://www.equinix.co.uk/newsroom/press-releases/pr/123467/equinix-agrees-to-divest-eight-european-assets-to-digital-realty-trust-inc/

51 https://www.ft.com/content/2c2c02f0-29b8-333a-af16-c8388d12aaa7

52 https://www.engineeringnews.co.za/article/teraco-raises-another-r158m-for-data-centre-expansion-in-sa-2011-06-02

53 https://investinganswers.com/dictionary/1/10-bagger

54 https://data-economy.com/data-economy-frontline-equinixs-new-2bn-expansion-plans-mean-big-business-for-all/

55 https://www.statista.com/statistics/802690/worldwide-connected-devices-by-access-technology/

Printed in Great Britain
by Amazon

45650359R00154